Felix Karbstein

Integrating out the Dirac Sea

Felix Karbstein

# Integrating out the Dirac Sea

## A no-sea effective theory approach

Südwestdeutscher Verlag für Hochschulschriften

**Impressum/Imprint (nur für Deutschland/ only for Germany)**
Bibliografische Information der Deutschen Nationalbibliothek: Die Deutsche Nationalbibliothek verzeichnet diese Publikation in der Deutschen Nationalbibliografie; detaillierte bibliografische Daten sind im Internet über http://dnb.d-nb.de abrufbar.

Alle in diesem Buch genannten Marken und Produktnamen unterliegen warenzeichen-, marken- oder patentrechtlichem Schutz bzw. sind Warenzeichen oder eingetragene Warenzeichen der jeweiligen Inhaber. Die Wiedergabe von Marken, Produktnamen, Gebrauchsnamen, Handelsnamen, Warenbezeichnungen u.s.w. in diesem Werk berechtigt auch ohne besondere Kennzeichnung nicht zu der Annahme, dass solche Namen im Sinne der Warenzeichen- und Markenschutzgesetzgebung als frei zu betrachten wären und daher von jedermann benutzt werden dürften.

Verlag: Südwestdeutscher Verlag für Hochschulschriften Aktiengesellschaft & Co. KG
Dudweiler Landstr. 99, 66123 Saarbrücken, Deutschland
Telefon +49 681 37 20 271-1, Telefax +49 681 37 20 271-0, Email: info@svh-verlag.de
Zugl.: Erlangen, FAU, Diss., 2009

Herstellung in Deutschland:
Schaltungsdienst Lange o.H.G., Zehrensdorfer Str. 11, D-12277 Berlin
Books on Demand GmbH, Gutenbergring 53, D-22848 Norderstedt
Reha GmbH, Dudweiler Landstr. 99, D- 66123 Saarbrücken
**ISBN: 978-3-8381-1010-3**


**Imprint (only for USA, GB)**
Bibliographic information published by the Deutsche Nationalbibliothek: The Deutsche Nationalbibliothek lists this publication in the Deutsche Nationalbibliografie; detailed bibliographic data are available in the Internet at http://dnb.d-nb.de.

Any brand names and product names mentioned in this book are subject to trademark, brand or patent protection and are trademarks or registered trademarks of their respective holders. The use of brand names, product names, common names, trade names, product descriptions etc. even without
a particular marking in this works is in no way to be construed to mean that such names may be regarded as unrestricted in respect of trademark and brand protection legislation and could thus be used by anyone.


Publisher:
Südwestdeutscher Verlag für Hochschulschriften Aktiengesellschaft & Co. KG
Dudweiler Landstr. 99, 66123 Saarbrücken, Germany
Phone +49 681 37 20 271-1, Fax +49 681 37 20 271-0, Email: info@svh-verlag.de

Copyright © 2008 Südwestdeutscher Verlag für Hochschulschriften Aktiengesellschaft & Co. KG and licensors
All rights reserved. Saarbrücken 2008

Produced in USA and UK by:
Lightning Source Inc., 1246 Heil Quaker Blvd., La Vergne, TN 37086, USA
Lightning Source UK Ltd., Chapter House, Pitfield, Kiln Farm, Milton Keynes, MK11 3LW, GB
BookSurge, 7290 B. Investment Drive, North Charleston, SC 29418, USA
**ISBN: 978-3-8381-1010-3**

# Contents

1 Introduction   3

2 The Gross-Neveu model family in 1+1 dimensions   7
    2.1 Construction and validation of no-sea effective theory . . . . . . . . . . . . . . 9
        2.1.1 Gross-Neveu model ($GN_2$) . . . . . . . . . . . . . . . . . . . . . . . . . 9
        2.1.2 Nambu–Jona-Lasinio model ($NJL_2$) . . . . . . . . . . . . . . . . . . . 20
    2.2 Application of no-sea effective theory - new results . . . . . . . . . . . . . . . 30
    2.3 Conclusions and outlook . . . . . . . . . . . . . . . . . . . . . . . . . . . . . . 35

3 The Walecka model   39
    3.1 Walecka model in 1+1 dimensions . . . . . . . . . . . . . . . . . . . . . . . . . 41
        3.1.1 Construction of no-sea effective theory . . . . . . . . . . . . . . . . . . 41
        3.1.2 Testing the no-sea effective theory . . . . . . . . . . . . . . . . . . . . . 50
        3.1.3 Application of no-sea effective theory . . . . . . . . . . . . . . . . . . . 52
    3.2 Walecka model in 3+1 dimensions . . . . . . . . . . . . . . . . . . . . . . . . . 55
    3.3 General observations and results . . . . . . . . . . . . . . . . . . . . . . . . . 61
        3.3.1 Uniform nuclear matter . . . . . . . . . . . . . . . . . . . . . . . . . . 61
        3.3.2 Spatially non-uniform systems . . . . . . . . . . . . . . . . . . . . . . . 65

4 Conclusions and outlook   69

A Gross-Neveu model family   72
    A.1 Results for coefficients defined in Sec. 2.2 . . . . . . . . . . . . . . . . . . . . . 72
    A.2 $GN_2$ model with counter term renormalization . . . . . . . . . . . . . . . . . 73

B Walecka model   77
    B.1 Walecka model with counter term renormalization . . . . . . . . . . . . . . . . 77
    B.2 Proof: " $-$ " loops with $\gamma^\nu$ insertions vanish . . . . . . . . . . . . . . . . . 79
    B.3 Vanishing effective interactions . . . . . . . . . . . . . . . . . . . . . . . . . . 81

B.4 Exact localized solutions of no-sea effective theory with 4-fermion-interactions . 82
B.5 Results for coefficients defined in Sec. 3.1.3 . . . . . . . . . . . . . . . . . . 84

# Chapter 1

# Introduction

Relativistic quantum field theory (QFT) constitutes an integral part of fundamental theoretical physics. Its most spectacular success is the standard model of elementary particle physics, incorporating quantum electrodynamics (QED), quantum chromodynamics (QCD) and the theory of weak interactions. The main dynamical ingredients of the standard model are interactions of elementary spin-$\frac{1}{2}$ fermions via the exchange of (gauge) bosons. Apart from this fundamental theory, effective field theories based on phenomenologically important, but non-fundamental degrees of freedom, adequate and valid in a certain energy range, are widely studied. One example is the Walecka model [1], mimicking basic features of nuclear physics, but starting from "elementary" nucleon and meson fields. The basic degrees of freedom in such effective theories, the nucleons, are again spin-$\frac{1}{2}$ fermions, interacting via the exchange of bosons. Note that in principle, nuclear physics should emerge as a direct manifestation of the standard model Lagrangian. However, in particular due to the complexity of QCD, practical "ab-initio" calculations are prohibitive. Finally, simple but exactly solvable model QFTs are intensely studied. Fermionic QFTs like the Gross-Neveu (GN) model family [2] are of particular interest here, as they share significant properties with QCD.

Fermions fulfill the Pauli exclusion principle [3] and hence obey Fermi-Dirac statistics [4, 5]. Unlike bosons, two identical fermions may not occupy the same quantum state simultaneously. In an attempt to reconcile quantum mechanics with special relativity, Dirac formulated the equation, which subsequently came to be known as Dirac equation [6]. Initially thought of as single particle equation, this equation provides a description of spin-$\frac{1}{2}$ particles and has several striking consequences. While in non-relativistic quantum mechanics the spin has to be put in by hand, in relativistic quantum mechanics it emerges as a direct manifestation of the Poincaré group. Moreover, the equation predicts an infinite set of quantum states with negative energy, which is obviously hard to interpret in terms of single particle states. Faced with this problem, Dirac subsequently adopted a multi-particle interpretation and resolved this puzzle by proposing that all negative energy states are completely filled in the vacuum, leading him also to the prediction of antiparticles as holes in the Dirac sea [7]. In this respect, the vacuum can be seen as an infinite sea of negative energy electrons ("Dirac sea").

Note however that despite its success, Dirac's interpretation is not fully satisfactory, as a truly relativistic theory has to account for the possibility of particle creation and annihilation. The residual difficulties have been overcome within the framework of QFT. Here the fields are quantized, i.e., they are expanded in their normal modes and single particle creation and annihilation operators. Hence, the fermion field occurring in the Dirac equation has to be reinterpreted as an operator in Fock space. The Dirac sea however remains and causes considerable difficulties in the explicit determination of physical observables in QFTs with fermions. In particular it gives rise to UV divergences. In general, the Dirac sea is not inert and affected by interaction effects. The presence of further (positive energy) fermions can polarize the Dirac sea. Turning to finite temperature, excited states of the Dirac sea can be formed, in which negative energy levels are depleted in favor of positive energy levels. Instead of filling all subsequent single particle energy levels, lower energy levels are depleted for higher ones.

One traditional way of addressing QFT is perturbation theory. However, this requires an adequate, small expansion parameter. While successful in QED, perturbation theory is of limited use in QCD where it is only applicable at large momentum transfers, i.e., in the realm of asymptotic freedom. The analysis of bound states for example requires a different approach. To study such questions, one nowadays often turns to numerical and non-perturbative techniques like lattice QCD. However, there are also approximations which, depending on the model under consideration, are capable of providing more physical insights. Here we are mainly concerned with mean-field-type approximations familiar from many-body physics. The basic idea of these approaches is to reduce the underlying interacting QFT to the problem of a single particle moving within the effective potential generated by all other particles. In this work, we turn to the simplest approach of this type, the relativistic Hartree approximation (RHA), and focus on questions which can be reasonably dealt with using the RHA. This allows us to address on the one hand "large $N$"-type model field theories where the RHA becomes exact and on the other hand popular, more phenomenological effective field theories like the Walecka model. Even in this simpler context, the Dirac sea has a highly non-trivial impact, significantly complicating any practical calculations. Note that in leading order of the large $N$ expansion, the relativistic Hartree-Fock (HF) approximation reduces to the RHA. In the context of "large $N$"-type model field theories one conventionally speaks of the HF approximation instead of the RHA.

We propose a novel approach to deal with the Dirac sea. Our aim is to integrate out the Dirac sea and to derive a "no-sea" effective theory featuring positive energy single particle states only. In this context we are interested in the general question, how a theory with positive energy states only emerges from the full underlying QFT. Furthermore, we ask for the manifestation of the Dirac sea within such an effective theory. This is related to the fundamental problem how non-relativistic many-body quantum mechanics arises from relativistic QFT. The other motivation has to do with applications. At least in certain parameter ranges, we hope to gain practical, calculational advantages, in particular in the determination of localized multi-fermion bound states. We do not try to develop our approach on an abstract, formal level, but rather within particular models.

One of the simplest interacting fermionic QFTs one could think of is the Gross-Neveu (GN) model [2], describing fermions interacting via a scalar four-fermion interaction in 1+1 spacetime dimensions. Surprisingly, introducing $N$ flavors and turning to the large $N$ limit, this model can be solved exactly. Moreover, it shows some interesting features, resembling basic properties of QCD, such as renormalizability, asymptotic freedom and without bare mass term, dimensional transmutation as well as chiral symmetry breaking. Since this particular model is well understood analytically, we use it to develop, test and validate our approach. Furthermore, a simple extension of the 1+1 dimensional GN model, namely adding a pseudoscalar four-fermion interaction, leads to the 1+1 dimensional Nambu Jona Lasino (NJL) model [8], which is less well understood than the GN model. Here our approach has a chance to provide new analytical insight.

Subsequently, we extend the methods developed in the context of the 1+1 GN model family to a QFT living in 3+1 dimensions. As the GN model is not renormalizable in 3+1 dimensions we look for a different quantum field theoretical model. A good candidate where mean-field-type approximations are believed to be adequate is the Walecka model [1], which has been intensely studied in the field of nuclear physics. The Walecka model can also be considered in 1+1 dimensions, where it shares basic features with the 1+1 GN model. We shall discuss the 1+1 dimensional Walecka model first. Having become acquainted with the model itself, we address the effect of increasing dimensions to 3+1 and discuss the Walecka model in 3+1 dimensions. We confront our findings with those obtained within the mean-field theory (MFT) approach, where the Dirac sea is simply ignored from the outset, and end with conclusions and an outlook.

# Chapter 2

# The Gross-Neveu model family in 1+1 dimensions

Exactly solvable models in quantum field theory which bear any resemblance to the real world are extremely rare. One example is provided by the Gross-Neveu model family [2], 1+1 dimensional four-fermion interaction models with $N$ flavors and Lagrangian

$$\mathcal{L} = \bar{\psi}\left(i\partial\!\!\!/ - m_0\right)\psi + \frac{g^2}{2}\left[(\bar{\psi}\psi)^2 + \lambda(\bar{\psi}i\gamma_5\psi)^2\right], \tag{2.1}$$

where $\lambda = 0$ or 1, and flavor indices are suppressed ($\bar{\psi}\psi = \sum_{k=1}^{N}\bar{\psi}_k\psi_k$ etc.). Depending on the value of $\lambda$, these models feature discrete ($\psi \to \gamma_5\psi$, $\lambda = 0$) or continuous ($\psi \to \exp(i\alpha\gamma_5)\psi$, $\lambda = 1$) chiral symmetry, possibly broken by the bare mass term $\sim m_0$. To avoid confusion we shall refer to the first model as Gross-Neveu ($GN_2$), the 2nd one as Nambu-Jona-Lasinio ($NJL_2$) model. From the point of view of strong interaction physics, they are most useful in the 't Hooft limit ($N \to \infty, Ng^2 = $ const.) to which we stick in the following. Rather than repeating all the well-known attractive features of these models, we refer the reader to some pertinent review articles [9, 10, 11] and the references therein.

Let us however emphasize two important, general relations that can be obtained directly from the Lagrangian. In order to get them, we turn to the Euler-Lagrange equations and derive the divergence of the (flavor neutral) vector current $j^\mu = \bar{\psi}\gamma^\mu\psi$ and axial vector current $j_5^\mu = \bar{\psi}\gamma^\mu\gamma_5\psi$,

$$\partial_\mu j^\mu = 0,$$
$$\partial_\mu j_5^\mu = 2\left(m_0 - (1-\lambda)g^2\bar{\psi}\psi\right)\bar{\psi}i\gamma_5\psi. \tag{2.2}$$

The first equation expresses conservation of fermion number in all models. The 2nd equation reflects the continuous chiral symmetry of the massless $NJL_2$ model (for $m_0 = 0, \lambda = 1$) and exhibits the source of chiral symmetry violation in those cases where the right hand side is non-zero. At $m_0 \neq 0, \lambda = 1$, it reduces to the standard PCAC relation. In 1+1 dimensions, vector and axial vector current are trivially related ($\gamma_5 = \gamma^0 \gamma^1$),

$$j_5^0 = j^1, \quad j_5^1 = j^0. \tag{2.3}$$

Taking an expectation value of Eqs. (2.2) in a static configuration (vacuum, multi-fermion bound state, dense matter) and using the factorization characteristic for the large $N$ limit together with Eqs. (2.3), we find

$$\partial_1 \langle j^1 \rangle = 0, \tag{2.4}$$

$$\partial_1 \langle j^0 \rangle = 2 \left( m_0 - (1 - \lambda) g^2 \langle \bar{\psi} \psi \rangle \right) \langle \bar{\psi} i \gamma_5 \psi \rangle. \tag{2.5}$$

Likewise we can take the thermal average of Eqs. (2.2) in the grand canonical ensemble at finite temperature and chemical potential. Relations (2.4) and (2.5) then remain valid provided that the expectation values are interpreted as thermal ones.

Whereas Eq. (2.4) apparently contains little dynamics, Eq. (2.5) turns out to be quite powerful. It relates the spatial derivative of the fermion density to the scalar and pseudoscalar condensates and the bare mass.

The Lagrangian (2.1) is known to possess multi-fermion bound states analogous to baryons or baryonium states in hadron physics. They were first constructed by Dashen, Hasslacher and Neveu (DHN) in the massless $GN_2$ model [12] and by Shei in the massless $NJL_2$ model [13]. The semi-classical method developed by these authors may be rephrased equivalently as a Dirac-Hartree-Fock calculation [14, 15]. Here, one solves the first-quantized, time-independent Dirac equation

$$\left( -\gamma_5 i \partial_x + \gamma^0 S + i \gamma^1 P \right) \psi_\alpha = E_\alpha \psi_\alpha \tag{2.6}$$

for single particle orbits with label $\alpha$ subject to self-consistency conditions for scalar and pseudoscalar mean fields,

$$S = m_0 - Ng^2 \sum_\alpha^{\text{occ}} \bar{\psi}_\alpha \psi_\alpha,$$

$$P = -\lambda Ng^2 \sum_\alpha^{\text{occ}} \bar{\psi}_\alpha i \gamma_5 \psi_\alpha. \tag{2.7}$$

A major challenge arises from the fact that the Dirac sea must be included in the sum over occupied states in (2.7). This problem was solved in Refs. [12, 13] by inverse scattering methods together with a careful subtraction of bound state and vacuum energies. At about the same time the proposal was made to solve these models "classically", neglecting the Dirac sea and including only the discrete valence level [16, 17]. This reduces the full problem to a non-linear Dirac equation, which has been solved in closed analytical form for both the $GN_2$ and $NJL_2$

models. It appeared that the classical fermionic solution was useful for the GN$_2$ model but completely failed in the NJL$_2$ case [13] for poorly understood reasons.

Here we reconsider the role of the Dirac sea in multi-fermion bound states of Gross-Neveu models. From a general field theoretic point of view, it should be possible to "integrate out" the Dirac sea and derive an effective Lagrangian for (positive energy) valence fermions only. In the large $N$ limit, it would then indeed be sufficient to solve the classical Euler-Lagrange equation of this effective theory for fermions. From such a point of view the above mentioned "no-sea" calculations may be re-interpreted as follows: The authors assumed that the effective Lagrangian is the same as the original Lagrangian (2.1), except that the bare parameters $m_0$ and $g^2$ are replaced by an effective mass and coupling constant. In the course of this work we will confront this implicit assumption with our results for an effective action derived from the underlying field theory.

Let us mention that although the solvable models, we concentrate on in this chapter, are restricted to 1+1 dimensions, the technique applied in deriving the effective Lagrangian is not. It is useful also in higher dimensional (mean-field-type) theories. This will become obvious in the next chapter, where the same approach will be applied to the Walecka model. The advantage of developing the methods in the context of Gross-Neveu models is the fact that the exact bound states are known analytically. This enables us to test our effective action quantitatively and make sure that the expansion is consistent to a given order in some small parameter. Since the derivation of $\mathcal{L}_{\text{eff}}$ is not completely straightforward due to the necessity of resummations, this has turned out to be quite helpful indeed.

## 2.1 Construction and validation of the no-sea effective theory

### 2.1.1 Gross-Neveu model (GN$_2$)

Our aim is to account for the effects of the Dirac sea by means of an effective theory of positive energy fermions only. Since we do not use the path integral but work canonically, we first have to explain what we mean by "integrating out the Dirac sea". We start with the massless GN$_2$ model described by the bare Lagrangian

$$\mathcal{L} = \bar{\psi} i \partial\!\!\!/ \psi + \frac{g^2}{2}(\bar{\psi}\psi)^2 \tag{2.8}$$

and focus on the DHN baryons. They are characterized by a discrete positive energy level occupied with $n \leq N$ fermions, all negative energy levels being completely filled. The presence of extra fermions polarizes the Dirac sea, distorting the single particle wave functions. Ideally,

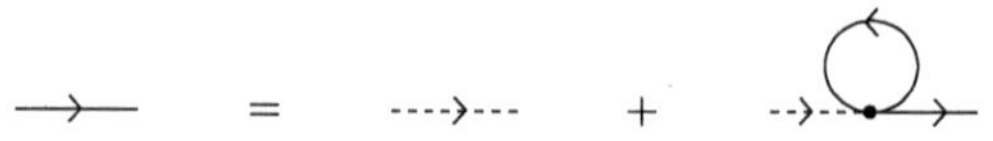

Figure 2.1: Dyson equation for fermion propagator in HF approximation. Dashed line: free propagator, solid line: dressed propagator.

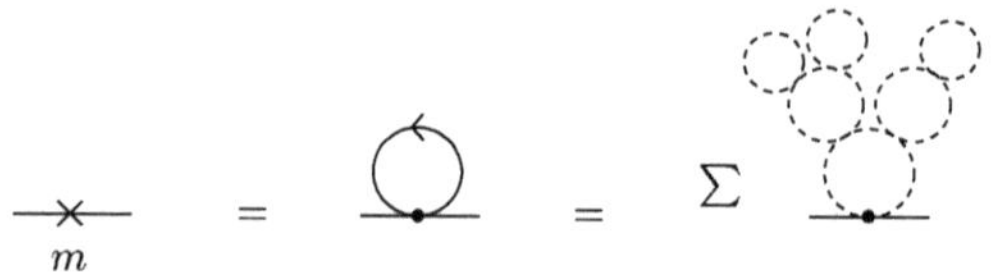

Figure 2.2: Dynamically generated fermion mass in the vacuum. A typical "cactus" diagram produced by an iterative solution of the HF equation in Fig. 2.1 is illustrated.

one would like to fully integrate out the Dirac sea. We have not been able to do this. Anyway, the full effective theory will invariably be non-local and therefore of less practical use. We therefore look for an expansion parameter which would enable us to derive an "almost local" effective action, consisting of a polynomial in the chiral condensate $\bar{\psi}\psi$ and its first few derivatives. From the DHN baryon we know that for small filling fraction $\nu = n/N$ of the valence level, the self-consistent scalar potential $S(x)$ differs from the physical (vacuum) fermion mass $m$ by a weak and slowly varying potential only. This suggests to use $\nu$ as expansion parameter and to restrict oneself to small filling fraction where the HF potential becomes soft.

The following derivation of the effective action is tailored to the HF approach to which we now turn. In the vacuum, a fermion mass is generated dynamically as shown graphically in Fig. 2.1. The self-consistent mass can be thought of as the sum over all one-particle-irreducible (1PI) cactus type diagrams, see Fig. 2.2. The (self-consistent) tadpole diagram receives contributions from all negative energy occupied states, leading to the gap equation

$$m\left(1 - \frac{Ng^2}{\pi}\ln\frac{\Lambda}{m}\right) = 0 \qquad (2.9)$$

(with $\Lambda/2$ as UV cutoff [14]). Clearly, since the physical fermion mass is a pure manifestation of the Dirac sea, it has to be put in by hand into an effective theory as an effective mass term

$$\mathcal{L}_{\text{eff}}^{(1)} = -m\bar{\psi}\psi. \qquad (2.10)$$

(From now on the superscript on $\mathcal{L}_{\text{eff}}$ refers to the number of loops of the self-energy diagram from which it was derived. Due to resummations, it may contain higher loop effects as well.)

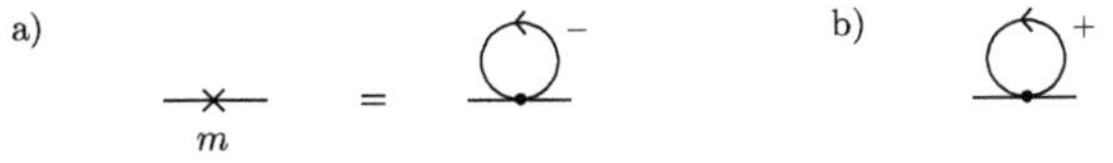

## 2.1. CONSTRUCTION AND VALIDATION OF NO-SEA EFFECTIVE THEORY

Figure 2.3: Decomposition of the fermion self-energy into a) negative and b) positive energy contributions.

Let us now turn to the problem of finite fermion number. The HF approach still has the same basic structure as in Fig. 2.1, except that the fermion self-energy is in general $x$-dependent. The propagator gets an extra contribution from the positive energy valence states. We denote the vacuum and valence particle contributions by " $-$ " and " $+$ ", respectively. For the free massive propagator for instance, the corresponding decomposition can be inferred from the well known result for finite temperature and chemical potential [18, 19],

$$\mathrm{i}G(p) = \mathrm{i}G_-(p) + \mathrm{i}G_+(p) ,$$

$$\mathrm{i}G_-(p) = \frac{\mathrm{i}}{\not{p} - m + \mathrm{i}\eta} ,$$

$$\mathrm{i}G_+(p) = -2\pi\delta(p^2 - m^2)(\not{p} + m)$$
$$\times \left[\theta(-p_0)\theta(E_f - E_p) + \theta(p_0)\theta(-E_f - E_p)\right] . \tag{2.11}$$

Here, $\mathrm{i}G_-(p)$ is just the free Feynman propagator, $E_f$ the Fermi energy or chemical potential. The one-loop self-energy (i.e., the tadpole) then naturally splits up into the two pieces shown in Fig. 2.3. In the effective theory with the Dirac sea integrated out, diagram a) is accounted for by the mass term (2.10). Diagram b) corresponds to the lowest order "no-sea" HF calculation with the original four-fermion interaction from Eq. (2.8). This splitting of the tadpole into $+/-$ pieces is also the key for the following systematic procedure. For a given number of loops, draw all topologically distinct 1PI cactus diagrams as shown in Fig. 2.4 up to five loops. Split each loop into a sum of " $+$ " and " $-$ " contributions. This generates $2^n$ distinct, "labeled" diagrams out of each $n$-loop diagram in Fig. 2.4. Then analyze each labeled diagram with respect to the question whether it is accounted for by the effective action with the Dirac sea integrated out. If it is not, add a new term to $\mathcal{L}_{\mathrm{eff}}$ which generates the corresponding fermion self-energy.

Although this scheme is straightforward and constructive, it is still too naive. We will see that it would be meaningless to terminate the procedure at any finite number of loops. Remember that the mass term (2.10) already includes an infinite number of diagrams due to the self-consistency condition (see Fig. 2.2). Otherwise, dimensional transmutation, a truly non-perturbative phenomenon, could not occur. We can only make sense out of the $n$-loop contribution to the effective Lagrangian if we perform appropriate resummations. This turns

12  CHAPTER 2. THE GROSS-NEVEU MODEL FAMILY IN 1+1 DIMENSIONS

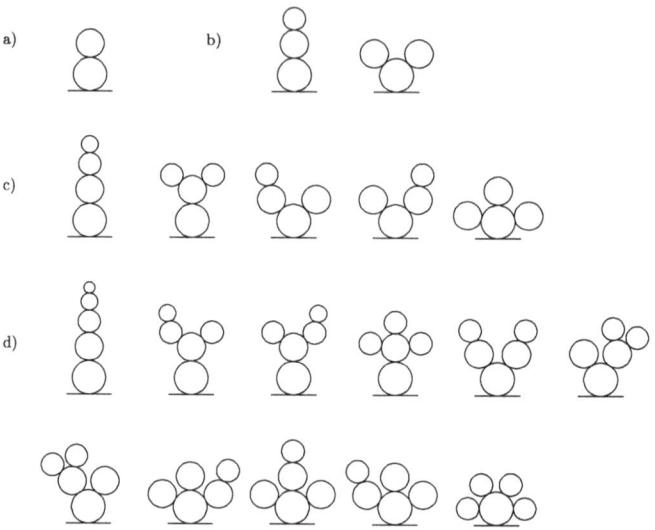

Figure 2.4: Topologically distinct diagrams contributing to the fermion self-energy in HF approximation from two up to five loops. They are needed to construct the effective action for the GN$_2$ model in NNLO, see the main text.

out to be a crucial element of the whole procedure and the way in which the bare coupling constant $g^2$ gets traded for a finite effective coupling constant.

To explain how this works, let us go through the first few terms of the loop expansion in some detail. Guided by the DHN baryon at small filling, we shall treat $\bar{\psi}\psi$ and $k^2$ (or $\Box$) as being of the same order $\epsilon$ and derive systematically all terms up to O($\epsilon^4$).

At two-loop order, the only diagram (out of 4 labeled diagrams) which induces a new term in $\mathcal{L}_{\text{eff}}$ is the one shown in Fig. 2.5a. This corresponds to a vertex correction to the " + " tadpole in Fig. 2.3b. Technically, it involves the scalar vacuum polarization loop (labeled " − " in Fig. 2.5a), where scalar means that the vertices are just 1. We evaluate it by standard Feynman rules and expand the result in powers of the 4-momentum $k$ flowing through the graph. The result for the sum of the " + " tadpole and Fig. 2.5a then gives the following self energy contribution,

$$\delta\Sigma(k) = -g^2 \langle\bar{\psi}\psi\rangle_k \left\{ 1 + \frac{Ng^2}{\pi} \left( \ln\frac{\Lambda}{m} - 1 + \frac{k^2}{12m^2} + \frac{(k^2)^2}{120m^4} + \mathrm{O}(k^6) \right) \right\}. \quad (2.12)$$

Here, the scalar condensate in momentum space, $\langle\bar{\psi}\psi\rangle_k$, is defined so as to include only " + " terms. Whereas the $k^2$-dependent correction terms are suppressed (remember that $k^2 \sim \epsilon$),

## 2.1. CONSTRUCTION AND VALIDATION OF NO-SEA EFFECTIVE THEORY

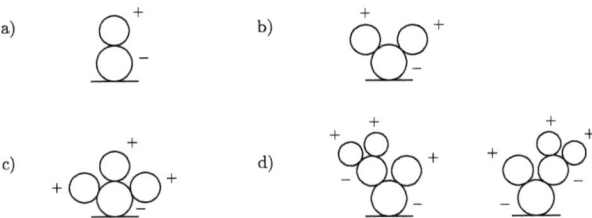

Figure 2.5: Labeled diagrams selected from the diagrams in Fig. 2.4, which are the seed for new terms in the effective Lagrangian. See main text for a detailed discussion.

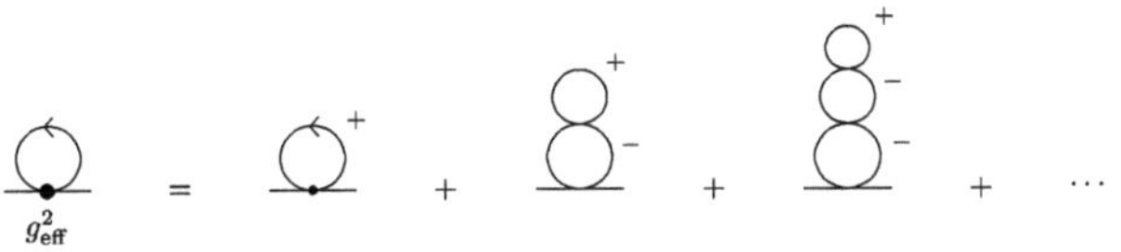

Figure 2.6: Resummation of scalar vacuum polarization graphs into a momentum dependent effective coupling. The infinite sum of " — " bubbles may be thought of as generating the $\sigma$ propagator.

the $k = 0$ contribution $\sim \ln(\Lambda/m)$ needs a special treatment. According to the gap equation (2.9), this term is actually of $O(1)$ so that the series expansion starting with Eq. (2.12) is meaningless as it stands. We therefore sum the whole series of graphs shown in Fig. 2.6 into a geometrical series, or, equivalently, replace the bare coupling constant by the effective, momentum dependent coupling

$$g_{\text{eff}}^2(k) = \frac{g^2}{1 - \frac{Ng^2}{\pi}\left(\ln\frac{\Lambda}{m} - 1 + \frac{k^2}{12m^2} + \frac{(k^2)^2}{120m^4}\right)}. \quad (2.13)$$

Due to the gap equation, the 1 in the denominator is now cancelled against $\frac{Ng^2}{\pi}\ln\left(\frac{\Lambda}{m}\right)$, and the bare coupling constant $g^2$ drops out of this expression. Expanding the $k$-dependent terms again [to $O(k^4)$], we arrive at the following momentum dependent effective coupling,

$$g_{\text{eff}}^2(k) = \frac{\pi}{N}\left(1 + \frac{k^2}{12m^2} + \frac{11(k^2)^2}{720m^4}\right). \quad (2.14)$$

It is now easy to write down a two-loop effective Lagrangian which would give just these correction terms (correct at two-loop level, but containing higher order terms due to resummation),

$$\mathcal{L}_{\text{eff}}^{(2)} = \frac{\pi}{2N}(\bar{\psi}\psi)^2 - \frac{\pi}{24Nm^2}(\Box\bar{\psi}\psi)(\bar{\psi}\psi) + \frac{11\pi}{1440Nm^4}(\Box^2\bar{\psi}\psi)(\bar{\psi}\psi). \quad (2.15)$$

Notice that the bare coupling constant $g^2$ in the original $(\bar{\psi}\psi)^2$ term has been replaced by $\pi/N$, and new four-fermion interaction terms containing derivatives of $\bar{\psi}\psi$ have been generated. Higher order terms could easily be obtained from the full $k$-dependence of the vacuum polarization graph if desired. The result is manifestly non-perturbative, as it does not contain $g^2$, and finite owing to the use of the gap equation.

It may be worthwhile to pause here and comment on the value of the leading order effective coupling, $g_{\text{eff}}^2 = \pi/N$ [see Eq. (2.14)]. This quantity already appeared in the original paper by Gross and Neveu [2] as fermion-fermion scattering amplitude at zero momentum. These authors also discuss the scalar $\sigma$ meson, pointing out that the square of the fermion-antifermion-$\sigma$ coupling constant is given by $g_{\sigma F\bar{F}}^2 = 4\pi m^2/N$. One can then understand the effective coupling in a similar way as in the old Fermi theory of weak interactions, namely as a product of two coupling constants and a heavy boson propagator ($M_\sigma = 2m$),

$$g_{\sigma F\bar{F}} \frac{1}{M_\sigma^2} g_{\sigma F\bar{F}} = \frac{\pi}{N}. \tag{2.16}$$

Keeping the momentum dependent terms approximately accounts for the finite range of the $\sigma$-exchange.

We now turn to the three-loop graphs. The two topologically distinct graphs in Fig. 2.4b give rise to $2 \times 2^3 = 16$ labeled subgraphs. By inspection we find that only the graph shown in Fig. 2.5b is the seed for a new term in the effective action, all other diagrams being generated by lower order terms or self-consistency in the "+" sector. Evaluating the fermion loop with three scalar vertices and finite momenta $k_1, k_2$ from the two "+" loops, we find the contribution to the self-energy

$$\delta\Sigma(k) = -\int \frac{dk_1}{2\pi} \frac{dk_2}{2\pi} (2\pi)\delta(k - k_1 - k_2)$$
$$\times \frac{(Ng^2)^3}{2\pi m} \left(1 + \frac{k_1^2 + k_2^2 + k_1 k_2}{6m^2}\right) \frac{\langle \bar{\psi}\psi\rangle_{k_1} \langle \bar{\psi}\psi\rangle_{k_2}}{N^2}. \tag{2.17}$$

Here, it is sufficient to go to $O(k^2)$. Resumming bubbles by replacing each of the three couplings $Ng^2$ at the vertices by effective ones analogously to Fig. 2.6,

$$(Ng^2)^3 \to Ng_{\text{eff}}^2(k_1) Ng_{\text{eff}}^2(k_2) Ng_{\text{eff}}^2(k_1 + k_2), \tag{2.18}$$

we find to $O(k^2)$

$$\delta\Sigma(k) = -\int \frac{dk_1}{2\pi} \frac{dk_2}{2\pi} (2\pi)\delta(k - k_1 - k_2)$$
$$\times \frac{\pi^2}{2m} \left(1 + \frac{k_1^2 + k_2^2 + k_1 k_2}{3m^2}\right) \frac{\langle \bar{\psi}\psi\rangle_{k_1} \langle \bar{\psi}\psi\rangle_{k_2}}{N^2}. \tag{2.19}$$

Figure 2.7: Illustration of the difference between the eight-fermion interactions $\mathcal{L}_{\text{eff}}^{(4)}$ and $\mathcal{L}_{\text{eff}}^{(5)}$ in Eqs. (2.21) and (2.22). The wiggly lines are $\sigma$ mesons and represent the bubble sum shown in Fig. 2.6. Each external line ends with a scalar density $\bar{\psi}\psi$.

The effective Lagrangian that yields this self-energy includes the following six-fermion interactions,

$$\mathcal{L}_{\text{eff}}^{(3)} = \frac{\pi^2}{6mN^2}(\bar{\psi}\psi)^3 - \frac{\pi^2}{12m^3N^2}(\Box\bar{\psi}\psi)(\bar{\psi}\psi)^2. \quad (2.20)$$

At four-loop order, there is again only a single graph (out of $5 \times 2^4 = 80$ labeled diagrams, see Fig. 2.4c) which generates a new term in the effective action. It is shown in Fig. 2.5c. The central loop labeled "$-$" with four scalar vertices is easy to compute, since we only need its value at $k = 0$. Resumming the bubbles by substituting $Ng^2 \to \pi$ [to $O(k^0)$], an eight-fermion interaction,

$$\mathcal{L}_{\text{eff}}^{(4)} = \frac{\pi^3}{24m^2N^3}(\bar{\psi}\psi)^4, \quad (2.21)$$

is induced. This is not yet the whole story to $O(\epsilon^4)$ though. Out of the $11 \times 2^5 = 352$ labeled five-loop diagrams derived from Fig. 2.4d, the two shown in Fig. 2.5d give rise to a further eight-fermion interaction term of the same order as (2.21). Applying the by now familiar resummation it is given by

$$\mathcal{L}_{\text{eff}}^{(5)} = \frac{\pi^3}{8m^2N^3}(\bar{\psi}\psi)^4. \quad (2.22)$$

The different origin of the contributions (2.21) and (2.22) becomes clearer if one interprets the scalar bubble sum as $\sigma$-meson propagator. While the four-loop contribution (2.21) corresponds to a $\sigma\sigma \to \sigma\sigma$ point-like interaction due to a heavy fermion loop, the five-loop term (2.22) has a different topology. It arises from the process $\sigma\sigma \to \sigma \to \sigma\sigma$ with an iterated effective three-$\sigma$ vertex, see Fig. 2.7.

This completes the discussion of all terms up to $O(\epsilon^4)$. We summarize by writing down the effective low-energy Lagrangian for the massless $GN_2$ model valid to $O(\epsilon^4)$ where the Dirac sea has been integrated out,

$$\mathcal{L}_{\text{eff}} = \bar{\psi}(i\slashed{\partial} - m)\psi + \frac{\pi}{2N}(\bar{\psi}\psi)^2 - \frac{\pi}{24m^2N}(\Box\bar{\psi}\psi)(\bar{\psi}\psi) + \frac{\pi^2}{6mN^2}(\bar{\psi}\psi)^3$$
$$+ \frac{11\pi}{1440m^4N}(\Box^2\bar{\psi}\psi)(\bar{\psi}\psi) - \frac{\pi^2}{12m^3N^2}(\Box\bar{\psi}\psi)(\bar{\psi}\psi)^2 + \frac{\pi^3}{6m^2N^3}(\bar{\psi}\psi)^4. \quad (2.23)$$

By counting powers of $\bar{\psi}\psi$ and $\Box$, we identify the first two terms as leading order (LO), the next two as next-to-leading order (NLO) and the remaining three as NNLO approximation. To LO, a mass term is generated and the coupling constant $g^2$ of the four-fermion interaction is replaced by the effective coupling constant $\pi/N$. In higher orders, the Dirac sea manifests itself through momentum dependent couplings and the appearance of six- and eight-fermion interactions, to the order we are working. Here we are in the fortunate position of being able to evaluate all coefficients of the effective Lagrangian from the underlying field theory. Technically, the calculation of the coefficients only involves standard one-loop Feynman diagrams in the vacuum, without any reference to finite fermion density or baryons. Nevertheless, the Lagrangian (2.23) should be adequate to predict properties of baryons or baryonic matter by means of a purely classical calculation.

We shall test this conjecture against known results for the full $GN_2$ model in two different ways. First, we do a "no-sea" HF calculation for matter at low density, assuming unbroken translational invariance. Second, we evaluate the baryon with small occupation of the valence level by solving a non-linear Dirac equation.

If the condensate $\langle \bar{\psi}\psi \rangle$ is assumed to be translationally invariant, only the non-derivative terms of the effective action enter, i.e.,

$$\mathcal{L}'_{\text{eff}} = \bar{\psi}(i\slashed{\partial} - m)\psi + \frac{\pi}{2N}(\bar{\psi}\psi)^2 + \frac{\pi^2}{6mN^2}(\bar{\psi}\psi)^3 + \frac{\pi^3}{6m^2N^3}(\bar{\psi}\psi)^4 . \quad (2.24)$$

The Euler-Lagrange equation allows us to identify the effective mass $M$ via

$$\frac{\partial}{\partial \bar{\psi}} \mathcal{L}'_{\text{eff}} = (i\slashed{\partial} - M)\psi = 0 \quad (2.25)$$

with

$$M = m - \frac{\pi}{N}\langle\bar{\psi}\psi\rangle - \frac{\pi^2}{2mN^2}\langle\bar{\psi}\psi\rangle^2 - \frac{2\pi^3}{3m^2N^3}\langle\bar{\psi}\psi\rangle^3 . \quad (2.26)$$

By construction, the condensate $\langle\bar{\psi}\psi\rangle$ that appears here refers only to the positive energy sector. Using the matter part $iG_+$ of the Dirac propagator (2.11) ($k_f = \sqrt{E_f^2 - M^2}$ denotes the Fermi momentum), we find

$$\langle\bar{\psi}\psi\rangle = N\frac{M}{\pi} \ln\left(\frac{k_f + \sqrt{k_f^2 + M^2}}{M}\right) . \quad (2.27)$$

A low density expansion in $k_f$ yields

$$M = m - k_f - \frac{(k_f)^2}{2m} - \frac{(k_f)^3}{2m^2} \pm \cdots \quad (2.28)$$

for the solution of Eqs. (2.26) and (2.27), in perfect agreement with the Taylor expansion of the exact result [9],

$$M = \sqrt{m^2 - 2mk_f} . \quad (2.29)$$

## 2.1. CONSTRUCTION AND VALIDATION OF NO-SEA EFFECTIVE THEORY

As is well known, this particular HF solution is not a minimum of the action. If translational invariance is assumed to be unbroken, there is a first order phase transition which requires a Maxwell construction, see Ref. [9]. Nevertheless this unphysical solution can be used as a purely algebraic test of the effective action. We thus confirm that the effects of the Dirac sea are encoded in the effective action, both through the mass and coupling constant of the four-fermion interaction and through induced many-fermion interaction terms. Obviously, the lower the density one is interested in, the smaller the number of terms that need to be kept in $\mathcal{L}_{\text{eff}}$.

A physically more relevant test case is the DHN baryon. We start with the Euler-Lagrange equation derived from the full effective Lagrangian (2.23) and everywhere replace $\bar{\psi}\psi$ by its ground state expectation value, this time determined by a single valence state. Let us define the $x$-dependent condensate

$$s = -\frac{\pi}{N}\langle\bar{\psi}\psi\rangle = -\pi\nu\bar{\psi}_0\psi_0 \qquad (2.30)$$

with $\nu = n/N$. Since $s(x)$ is time independent, we get

$$\Box\langle\bar{\psi}\psi\rangle = \frac{N}{\pi}s''. \qquad (2.31)$$

Consequently, the Euler-Lagrange equation can then be cast into the form of the following non-linear Dirac equation for the valence level,

$$\left(-\gamma_5 i\partial_x + \gamma^0 S\right)\psi_0 = E_0\psi_0, \qquad (2.32)$$

where the scalar potential $S$ is given by

$$S = m + s + \frac{1}{12m^2}s'' - \frac{1}{2m}s^2 + \frac{11}{720m^4}s^{IV} - \frac{1}{6m^3}\left[2s''s + (s')^2\right] + \frac{2}{3m^2}s^3, \qquad (2.33)$$

and $s$ in turn depends on $\psi_0$ through Eq. (2.30). Eqs. (2.32) and (2.33) have to be solved self-consistently subject to the normalization condition

$$\int \mathrm{d}x\, \psi_0^\dagger \psi_0 = 1. \qquad (2.34)$$

The mass of the baryon in the effective theory is then calculated as follows: Deduce the Hamiltonian density from $\mathcal{L}_{\text{eff}}$ (2.23) in the standard way,

$$\mathcal{H}_{\text{eff}} = \frac{\partial \mathcal{L}_{\text{eff}}}{\partial \dot{\psi}}\dot{\psi} - \mathcal{L}_{\text{eff}}. \qquad (2.35)$$

Since $\mathcal{L}_{\text{eff}}$ is linear in $\dot{\psi}$, this amounts to dropping the term containing the time derivative of $\psi$ and reverting the overall sign of $\mathcal{L}_{\text{eff}}$. The baryon mass is now simply given by

$$M_{\text{B}} = \int \mathrm{d}x\, \mathcal{H}_{\text{eff}}. \qquad (2.36)$$

To solve the "no-sea" Dirac-HF equation is not easy and would have to be done numerically in general. Since we know the exact answer for $\psi_0$ from the full $GN_2$ model and are primarily interested in checking our effective action, we proceed differently. We use the following representation of the $\gamma$ matrices,

$$\gamma^0 = -\sigma_1, \quad \gamma^1 = i\sigma_3, \quad \gamma_5 = \gamma^0\gamma^1 = -\sigma_2. \tag{2.37}$$

The (positive energy) valence level of the DHN baryon then has the spinor wave function and energy [14]

$$\psi_0(x) = \frac{\sqrt{ym}}{2}\begin{pmatrix} \frac{1}{\cosh\xi_-} \\ -\frac{1}{\cosh\xi_+} \end{pmatrix}, \quad E_0 = m\sqrt{1-y^2}, \tag{2.38}$$

with the definitions

$$\xi_\pm = ymx \pm \frac{1}{2}\operatorname{artanh} y, \quad y = \sin\theta, \quad \theta = \frac{\pi\nu}{2}. \tag{2.39}$$

We plug this ansatz into the non-linear Dirac equation (2.32), expand in the filling fraction $\nu$ and check with MAPLE that the equation is indeed satisfied up to (including) $O(\nu^6)$. For the baryon mass (2.36) we obtain at this order

$$M_B = nm\left(1 - \frac{\pi^2\nu^2}{24} + \frac{\pi^4\nu^4}{1920} - \frac{\pi^6\nu^6}{322560}\right). \tag{2.40}$$

We have truncated the series because higher order terms are not reliable, as they would require an improved effective action. The exact DHN baryon mass is given by the compact formula

$$M_B = nm\frac{\sin\theta}{\theta} \tag{2.41}$$

with $\theta$ as defined in Eq. (2.39). If we expand this function in powers of $\nu$, the series indeed starts with (2.40). This is obviously a very good test of our effective action and shows that the terms retained as well as the resummations done are consistent.

In the massless $GN_2$ model, the only regime where an almost local effective action for the baryon can be justified is $\nu \ll 1$, i.e., small valence filling fraction. Once we include a bare mass term, there is yet another handle to suppress Dirac sea effects and make the scalar potential softer, namely by increasing the bare mass. This incites us to generalize the "no-sea" effective action to the massive $GN_2$ model. The only difference to the previous calculation comes from the modified gap equation which now reads [20]

$$\frac{\pi}{Ng^2} = \gamma + \ln\frac{\Lambda}{m}, \quad \gamma := \frac{\pi}{Ng^2}\frac{m_0}{m}. \tag{2.42}$$

While the diagrams used to derive the effective action are the same as above, the algebra slightly changes. We immediately jump to the final effective Lagrangian for the massive $GN_2$

## 2.1. CONSTRUCTION AND VALIDATION OF NO-SEA EFFECTIVE THEORY

model in NNLO,

$$\mathcal{L}_{\text{eff}} = \bar{\psi}\left(i\slashed{\partial} - m\right)\psi + \frac{\pi}{2N}\frac{1}{(1+\gamma)}(\bar{\psi}\psi)^2 - \frac{\pi}{24m^2N}\frac{1}{(1+\gamma)^2}(\Box\bar{\psi}\psi)(\bar{\psi}\psi)$$
$$+ \frac{\pi^2}{6mN^2}\frac{1}{(1+\gamma)^3}(\bar{\psi}\psi)^3 + \frac{11\pi}{1440m^4N}\frac{(1+6\gamma/11)}{(1+\gamma)^3}(\Box^2\bar{\psi}\psi)(\bar{\psi}\psi)$$
$$- \frac{\pi^2}{12m^3N^2}\frac{(1+\gamma/2)}{(1+\gamma)^4}(\Box\bar{\psi}\psi)(\bar{\psi}\psi)^2 + \frac{\pi^3}{6m^2N^3}\frac{(1+\gamma/4)}{(1+\gamma)^5}(\bar{\psi}\psi)^4. \quad (2.43)$$

It reduces to Eq. (2.23) in the chiral limit or, equivalently, at $\gamma = 0$. In order to test our effective Lagrangian, we turn to the known exact baryons of the massive GN$_2$ model [21, 22, 23]. Defining $s(x)$ as in Eq. (2.30), the non-linear Dirac equation (2.32) now contains the scalar potential

$$S = m + \frac{1}{(1+\gamma)}s + \frac{1}{12m^2}\frac{1}{(1+\gamma)^2}s'' - \frac{1}{2m}\frac{1}{(1+\gamma)^3}s^2 + \frac{11}{720m^4}\frac{(1+6\gamma/11)}{(1+\gamma)^3}s^{IV}$$
$$- \frac{1}{6m^3}\frac{(1+\gamma/2)}{(1+\gamma)^4}\left[2s''s + (s')^2\right] + \frac{2}{3m^2}\frac{(1+\gamma/4)}{(1+\gamma)^5}s^3. \quad (2.44)$$

The exact spinor wave function and the energy remain the same as in Eqs. (2.38), but the relationship between the parameter $y$ and the valence occupation fraction $\nu$ changes to

$$\frac{\nu}{2} = \frac{\theta}{\pi} + \frac{\gamma}{\pi}\tan\theta, \quad y = \sin\theta. \quad (2.45)$$

Eq. (2.45) can be solved for $\theta$ by means of a power series expansion in $\nu$ for given $\gamma$. The exact baryon mass is given by

$$M_B = \frac{2mN}{\pi}\left[\sin\theta + \gamma\,\text{artanh}\,(\sin\theta)\right]. \quad (2.46)$$

We have verified with MAPLE that the full wave function $\psi_0$ solves the non-linear Dirac equation derived from Eq. (2.43) to an accuracy of $O(\nu^6)$, as in the massless case. The baryon mass calculated from the effective action is found analytically to be

$$M_B = nm\left(1 - \frac{\pi^2\nu^2}{24(1+\gamma)^2} + \frac{\pi^4(1+9\gamma)\nu^4}{1920(1+\gamma)^5} - \frac{\pi^6(1 - 54\gamma + 225\gamma^2)\nu^6}{322560(1+\gamma)^8}\right), \quad (2.47)$$

the generalization of Eq. (2.40) to arbitrary $\gamma$. To the given order this agrees once again with the result based upon the series expansion of the exact equations (2.45) and (2.46).

These results show that with our "no-sea" effective action, we have indeed achieved what we were aiming at. Depending on the number of terms one is willing to include, one can systematically get an increasingly accurate value for the baryon mass and valence wave function. Note that the expansion parameter changes from $\nu$ in the chiral limit to $\nu/\gamma$ in the heavy quark limit. As a matter of fact, the truncated series (2.47) is a much better guide to the exact baryon mass in the full range of the parameters $(\nu, \gamma)$ than could have been expected on the basis of our derivation. We find that the NNLO effective action yields the binding energy of the baryon with an error of less than 1.5% even for full occupation ($\nu = 1$) and arbitrary $\gamma$. In the chiral limit, the error is below 0.3% for all values of $\nu$. Even the LO effective action yields results no worse than 10-20% for the binding energy.

The most important result of this section is the effective Lagrangian (2.43) for the massive $GN_2$ model. It contains the corresponding Lagrangian (2.23) for the massless $GN_2$ model as a special case ($\gamma = 0$). The applications to homogeneous matter and baryons show that our general scheme is correct and underline the systematic character of the expansion. In the derivation, we only had to compute standard one-loop Feynman diagrams with increasing number of external lines. Nevertheless, we demonstrated that an excellent approximation to the baryon mass can be obtained even in regions of the parameter space where we had no a-priori reason to assume that our truncation of the effective action was justified.

### 2.1.2 Nambu–Jona-Lasinio model ($NJL_2$)

The Lagrangian of the Gross-Neveu model with continuous chiral symmetry ($NJL_2$ model) is

$$\mathcal{L} = \bar{\psi} i \partial\!\!\!/ \psi + \frac{g^2}{2} \left[ (\bar{\psi}\psi)^2 + (\bar{\psi} i \gamma_5 \psi)^2 \right]. \qquad (2.48)$$

In this model, the Dirac sea has a more dramatic impact on the hadrons of the theory than in the case of the $GN_2$ model. This is already clear from the existence of massless mesons and baryons, despite the fact that the elementary fermions acquire a dynamical mass [9]. As compared to the GN model, one has to expect additional complications when integrating out the Dirac sea. In the present section, we therefore restrict ourselves to the LO calculation of the effective action. It turns out that the LO calculation in the $NJL_2$ model is about as complex as the NNLO calculation in the $GN_2$ model described in Sect. II.

The general procedure and the diagrams to be considered stay the same as before, except that each vertex can now be either 1 (scalar) or $i\gamma_5$ (pseudoscalar). Some diagrams vanish due to parity selection rules.

At one-loop order (tadpole), the scalar contribution yields an effective mass just like in the $GN_2$ model. The pseudoscalar tadpole vanishes, so that the one-loop contribution to $\mathcal{L}_{\text{eff}}$ is again a standard mass term,

$$\mathcal{L}_{\text{eff}}^{(1)} = -m\bar{\psi}\psi. \qquad (2.49)$$

At two-loop level, we have to consider that the two vertices in Fig. 2.5a can both be either scalar or pseudoscalar. The first case is identical to the $GN_2$ model and, after resumming the scalar bubbles, yields the first term in Eq. (2.15) (higher order terms are discarded since we

## 2.1. CONSTRUCTION AND VALIDATION OF NO-SEA EFFECTIVE THEORY

now work to LO only). The second case involves a pseudoscalar vacuum polarization. For the sum of the " $+$ " tadpole and this correction, we obtain to $O(k^2)$

$$\delta\Sigma(k) = -g^2 \langle \bar{\psi} i\gamma_5 \psi \rangle_k \left\{ 1 + \frac{Ng^2}{\pi} \left( \ln\frac{\Lambda}{m} + \frac{k^2}{4m^2} \right) \right\}. \tag{2.50}$$

This should be compared to the corresponding scalar result, Eq. (2.12). Using the same arguments for resumming the inner " $-$ " bubbles as in the scalar case, we get (denoting the pseudoscalar coupling by $G_{\text{eff}}$)

$$G_{\text{eff}}^2(k) = \frac{g^2}{1 - \frac{Ng^2}{\pi}\left(\ln\frac{\Lambda}{m} + \frac{k^2}{4m^2}\right)}. \tag{2.51}$$

Invoking the gap equation (2.9) now yields the result

$$G_{\text{eff}}^2(k) = -\frac{4m^2}{k^2}\frac{\pi}{N}. \tag{2.52}$$

The pseudoscalar bubble sum has produced the pole of the massless "pion". Actually, the difference between (2.52) and the leading order result for the scalar coupling $g_{\text{eff}}^2 = \pi/N$ can easily be understood: As already noted by Gross and Neveu [2], the coupling constants $g_{\pi F\bar{F}}$ and $g_{\sigma F\bar{F}}$ are identical due to chiral symmetry. In Eq. (2.16), we have interpreted the scalar effective coupling constant in terms of a $\sigma$-exchange, replacing the propagator by a constant, namely its value at $k^2 = 0$. However, this cannot be done in the case of the $\pi$-exchange. The limit $k^2 \to 0$ is singular and we must keep the leading momentum dependence,

$$g_{\pi F\bar{F}} \frac{1}{-k^2} g_{\pi F\bar{F}} = -\frac{4m^2}{k^2}\frac{\pi}{N}, \tag{2.53}$$

thus reproducing Eq. (2.52). Clearly, the pole in the effective pseudoscalar interaction has drastic consequences for the structure of the effective theory. Since the expansion around $k = 0$ is singular in the NJL$_2$ model, we have to reconsider the ordering principle behind our approach. Keeping the momentum dependence from the pion pole in Eq. (2.52), the total two-loop contribution to the effective Lagrangian assumes the non-local form

$$\mathcal{L}_{\text{eff}}^{(2)} = \frac{\pi}{2N}(\bar{\psi}\psi)^2 + \frac{2\pi m^2}{N}\bar{\psi}i\gamma_5\psi\frac{1}{\Box}\bar{\psi}i\gamma_5\psi. \tag{2.54}$$

As before we shall treat $\bar{\psi}\psi$ and $\Box$ as being of $O(\epsilon)$. We would like to test our effective action against the multi-fermion bound states of Shei, in analogy to our test against the DHN baryon in the GN$_2$ case. From these known bound states we infer that $\bar{\psi}i\gamma_5\psi$ is of $O(\epsilon^{3/2})$. Hence both terms in Eq. (2.54) are of the same $O(\epsilon^2)$, which defines our LO approximation.

At three-loop level we must again consider the graph in Fig. 2.5b. If all three couplings are scalar, it is of $O(\epsilon^3)$ and therefore NLO. We neglect it in the present case. The only other

Figure 2.8: Dressed pion propagator (dashed line), connected to an arbitrary number of scalar densities $\bar{\psi}\psi$ via $\sigma$-exchanges (wiggly lines). See Eqs. (2.56) and (2.57).

non-vanishing contribution involves one scalar and two pseudoscalar couplings. We evaluate the corresponding Feynman graph and resum the scalar and pseudoscalar bubbles into effective couplings, generating two pion poles. The resulting effective Lagrangian reads

$$\mathcal{L}_{\text{eff}}^{(3)} = \frac{8\pi^2 m^3}{N^2} \bar{\psi} i\gamma_5 \psi \frac{1}{\Box} \bar{\psi}\psi \frac{1}{\Box} \bar{\psi} i\gamma_5 \psi \,. \tag{2.55}$$

Notice that this is again of $O(\epsilon^2)$. Due to the inverse powers of momenta or $\Box$, unlike in the GN$_2$ case, there is no suppression of the three-loop contribution as compared to the two-loop contribution by a power of $\epsilon$. Even worse, one can identify a whole class of higher terms with arbitrary many loops contributing to the same order of $\epsilon$. This calls once again for a resummation, namely [add up the 2nd term in Eq. (2.54) and Eq. (2.55)]

$$\frac{2\pi m^2}{N} \bar{\psi} i\gamma_5 \psi \frac{1}{\Box} \left(1 + \frac{4\pi m}{N} \bar{\psi}\psi \frac{1}{\Box} + \cdots \right) \bar{\psi} i\gamma_5 \psi \to \frac{2\pi m^2}{N} \bar{\psi} i\gamma_5 \psi \frac{1}{\Box - \frac{4\pi m}{N} \bar{\psi}\psi} \bar{\psi} i\gamma_5 \psi \,. \tag{2.56}$$

We have summed up a special class of higher order diagrams, singled out by being of $O(\epsilon^2)$. They can be described as follows: Draw a string of pseudoscalar bubbles between two $\bar{\psi} i\gamma_5 \psi$ (valence) condensates. Then attach at most one scalar tadpole to any of the pseudoscalar loops (on either side). All diagrams generated with this prescription are summed up. If there are $n$ scalar condensates, this must be matched by $n+1$ pion poles. It is easy to convince oneself that any other way of attaching tadpoles to the bubble graphs is punished by a higher order in $\epsilon$. As a result of this discussion, we replace the effective action (2.54) induced by the two-loop graph with

$$\mathcal{L}_{\text{eff}}^{(2)} = \frac{\pi}{2N} (\bar{\psi}\psi)^2 + \frac{2\pi m^2}{N} \bar{\psi} i\gamma_5 \psi \frac{1}{\Box - \frac{4\pi m}{N} \bar{\psi}\psi} \bar{\psi} i\gamma_5 \psi \,. \tag{2.57}$$

The massless pion propagator has been superseded by the propagator of the pion in a scalar background potential and sums up higher loop contributions of $O(\epsilon^2)$ in $\mathcal{L}_{\text{eff}}$ as shown schematically in Fig. 2.8. Notice that the pion self-energy appearing in the denominator of Eq. (2.57) has a form reminiscent of the Gell-Mann, Oakes, Renner (GOR) relation [24] which relates the pion mass to the bare fermion mass and the chiral condensate,

$$m_\pi^2 = -\frac{4\pi m_0}{N} \langle \bar{\psi}\psi \rangle \,. \tag{2.58}$$

## 2.1. CONSTRUCTION AND VALIDATION OF NO-SEA EFFECTIVE THEORY

However, in Eq. (2.57), $m$ is the physical fermion mass, and $\bar\psi\psi$ refers to the ($x$-dependent) valence contribution to the condensate only. Nevertheless, the formal similarity is striking and may point to a deeper physics reason behind our rather technical resummation.

Due to the troublesome occurrence of inverse powers of $\epsilon$ in the pseudoscalar effective coupling, we have to go to even higher order in the loop expansion in order to identify all LO terms. At four-loop order, we had to consider Fig. 2.5c for the $GN_2$ model (where it was NNLO). If all four vertices are scalar (like in the $GN_2$ model), it is indeed of $O(\epsilon^4)$ and can be discarded for our present purpose. If two vertices are scalar and two are pseudoscalar, after resummation it will be of $O(\epsilon^3)$ and hence still negligible. The only $O(\epsilon^2)$ contribution is the diagram in Fig. 2.5c with four $i\gamma_5$ vertices. The four pion pole terms together with four pseudoscalar condensates conspire to give $O(\epsilon^2)$. The Feynman diagram calculation yields the effective action

$$\mathcal{L}^{(4)}_{\text{eff}} = -\frac{32\pi^3 m^6}{N^3}\left(\frac{1}{\Box}\bar\psi i\gamma_5\psi\right)^4. \tag{2.59}$$

According to our previous treatment of the two-loop diagram, we should once again replace every denominator $\Box$ by $\Box - \frac{4\pi m}{N}\bar\psi\psi$, thereby summing up another infinite set of diagrams of $O(\epsilon^2)$ but containing multi-fermion interactions of arbitrary order,

$$\mathcal{L}^{(4)}_{\text{eff}} = -\frac{32\pi^3 m^6}{N^3}\left(\frac{1}{\Box - \frac{4\pi m}{N}\bar\psi\psi}\bar\psi i\gamma_5\psi\right)^4. \tag{2.60}$$

Finally we turn to the five-loop graph in Fig. 2.5d, which contributed in NNLO in the $GN_2$ case. The interesting case is the one where there are four pseudoscalar vertices and one scalar vertex, the latter connecting the two " $-$ " bubbles. The calculation including all necessary resummations yields

$$\mathcal{L}^{(5)}_{\text{eff}} = \frac{32\pi^3 m^6}{N^3}\left(\frac{1}{\Box - \frac{4\pi m}{N}\bar\psi\psi}\bar\psi i\gamma_5\psi\right)^4. \tag{2.61}$$

It yields exactly the same magnitude as (2.60) but the opposite sign, so that the four- and five-loop terms cancel in the $NJL_2$ model. In our opinion this cancellation is an expression of the low energy theorem according to which the $\pi\pi$ interaction at zero momentum should vanish [25]. Here this comes about as a result of a cancellation between the $4\pi$ vertex and the process $\pi\pi \to \sigma \to \pi\pi$ involving the $\pi\pi\sigma$ vertex, see Fig. 2.9. It is interesting that the $NJL_2$ model respects the low energy theorems in the large $N$ limit, in spite of the usual strong reservations against Goldstone bosons in two dimensions [26].

We have now identified all terms of $O(\epsilon^2)$ in the effective Lagrangian. As announced, the calculation was more involved than in the $GN_2$ model. Whereas in the $GN_2$ case we had to deal with four-, six- and eight-fermion interactions in LO, NLO and NNLO, here we had to sum up terms corresponding to $2n$-fermion interactions with arbitrary $n$ already in LO. The

Figure 2.9: a) Effective direct $4\pi$ interaction corresponding to Eq. (2.60). b) Effective $4\pi$ interaction via $\sigma$-exchange leading to Eq. (2.61). The two diagrams cancel exactly due to the vanishing of the $\pi\pi$ interaction at zero momentum, in accordance with low energy theorems.

Figure 2.10: a) Effective fermion-fermion interaction through dressed $\pi$-exchange, giving rise to the non-local term in $\mathcal{L}_{\text{eff}}$, Eq. (2.62). b) Effective fermion-fermion interaction through $\sigma$-exchange, yielding the local $(\bar\psi\psi)^2$-term in $\mathcal{L}_{\text{eff}}$.

reason is the massless pion pole which induces inverse powers of $\epsilon$. It is responsible for the non-locality of the LO effective interaction for the NJL$_2$ model which nevertheless has a rather simple final form,

$$\mathcal{L}_{\text{eff}} = \bar\psi(i\slashed\partial - m)\psi + \frac{\pi}{2N}(\bar\psi\psi)^2 + \frac{2\pi m^2}{N}\bar\psi i\gamma_5\psi \frac{1}{\Box - \frac{4\pi m}{N}\bar\psi\psi}\bar\psi i\gamma_5\psi. \qquad (2.62)$$

The pseudoscalar term is associated with the fermion-fermion interaction through $\pi$-exchange, see Fig. 2.10a. The pion propagator has a long range and is dressed as in Fig. 2.8. The scalar term arises from $\sigma$-exchange, Fig. 2.10b, and is of zero range in LO. It is common to the GN$_2$ and NJL$_2$ models.

At first glance, the non-locality of $\mathcal{L}_{\text{eff}}$ is irritating, since it is expected to make the solution of the Euler-Lagrange equation much harder. However in the present case, the particular structure of the non-locality allows us to trade the non-local effective Lagrangian for a local one, at the cost of introducing an elementary pseudoscalar field $\Pi(x)$. The local effective Lagrangian equivalent to (2.62) is given by

$$\mathcal{L}_{\text{eff}}^{\text{loc}} = \bar\psi(i\slashed\partial - m)\psi + \frac{\pi}{2N}(\bar\psi\psi)^2 + \frac{1}{2}\partial_\mu\Pi\partial^\mu\Pi + \sqrt{\frac{4\pi}{N}}m\bar\psi i\gamma_5\psi\Pi + \frac{2\pi m}{N}\bar\psi\psi\Pi^2. \qquad (2.63)$$

Since we are in the large $N$ limit, $\Pi$ can be treated as a classical field. The equation of motion for $\Pi$ following from (2.63) is

$$\left(\Box - \frac{4\pi m}{N}\bar\psi\psi\right)\Pi = \sqrt{\frac{4\pi}{N}}m\bar\psi i\gamma_5\psi. \qquad (2.64)$$

## 2.1. CONSTRUCTION AND VALIDATION OF NO-SEA EFFECTIVE THEORY

Upon solving this inhomogeneous equation formally and plugging the result into Eq. (2.63) we indeed recover the non-local effective Lagrangian (2.62). So we wind up with a local field theory containing both fermions and an elementary $\pi$-meson, the latter described by the field $\Pi$.

Since the derivation was quite complicated (in particular identifying all LO terms), it is again crucial to test $\mathcal{L}_{\text{eff}}$ against exact results for the NJL$_2$ model. For this purpose, we choose the multi-fermion bound states of Shei [13]. They are analogous to the DHN baryon of the GN$_2$ model in the way they were derived (inverse scattering theory), but carry vanishing baryon number and should be regarded as "baryonium" states [27, 15].

Our strategy is as follows. Take the valence spinor wave function of Shei in the notation of Ref. [15],

$$\psi_0 = \sqrt{\frac{m|\cos\theta|}{2}} \frac{1}{\cosh\xi} \begin{pmatrix} \cos\theta/2 \\ \sin\theta/2 \end{pmatrix} \tag{2.65}$$

with

$$\xi = mx\cos\theta,$$
$$\theta = \left(\frac{3}{2} - \nu\right)\pi \tag{2.66}$$

and energy $E_0 = -m\sin\theta$. This yields the (valence) condensates

$$\langle\bar{\psi}\psi\rangle = Nm\nu \frac{\sin\pi\nu, \cos\pi\nu}{2\cosh^2\xi},$$
$$\langle\bar{\psi}i\gamma_5\psi\rangle = Nm\nu \frac{\sin^2\pi\nu}{2\cosh^2\xi}. \tag{2.67}$$

Eq. (2.64) for $\Pi$ becomes

$$\left(\partial_\xi^2 + \frac{2\pi\nu\cot\pi\nu}{\cosh^2\xi}\right)\Pi = -\frac{\nu\sqrt{\pi N}}{\cosh^2\xi} \tag{2.68}$$

and, to leading order in $\nu$, is solved by

$$\Pi = -\frac{\nu\sqrt{\pi N}}{1 + e^{2\xi}}. \tag{2.69}$$

The Euler-Lagrange equation for $\psi_0$ is equivalent to the "no-sea" Dirac-HF equation,

$$\left(\gamma_5 \frac{1}{i}\partial_x + \gamma^0 S(x) + i\gamma^1 P(x)\right)\psi_0 = E_0\psi_0, \tag{2.70}$$

with scalar and pseudoscalar potentials

$$S(x) = m - \frac{\pi}{N}\langle\bar{\psi}\psi\rangle - \frac{2\pi m}{N}\Pi^2 = m\left(1 - \frac{2\pi^2\nu^2}{1 + e^{2\xi}}\right),$$
$$P(x) = -\sqrt{\frac{4\pi}{N}}m\Pi = m\frac{2\pi\nu}{1 + e^{2\xi}}. \tag{2.71}$$

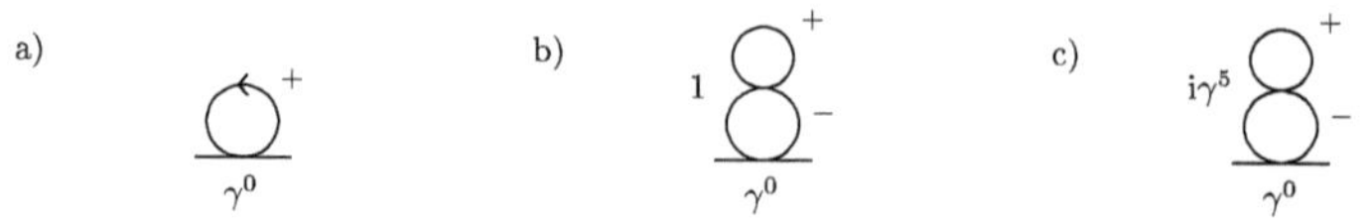

Figure 2.11: Illustration of the effective fermion density, Eq. (2.75). a) Valence contribution. b) Vanishing sea contribution due to the scalar interaction. c) Induced fermion density due to the pseudoscalar interaction.

We have kept the LO terms only. We find that $\psi_0$ satisfies the HF equation up to corrections of $O(\nu^3)$. The Hamiltonian density is given by

$$\mathcal{H} = \psi^\dagger \left(-\gamma_5 i\partial_x + \gamma^0 m\right) \psi - \frac{\pi}{2N}(\bar{\psi}\psi)^2 + \frac{1}{2}(\partial_x \Pi)^2 - \sqrt{\frac{4\pi}{N}} m\bar{\psi} i\gamma_5 \psi \Pi - \frac{2\pi m}{N}\bar{\psi}\psi\Pi^2. \quad (2.72)$$

Taking its expectation value in the valence state, we insert the above results for $\psi_0$ and $\Pi$ and integrate over $x$. The derivative term gives zero, the mass term has to be treated in NLO in $\nu$ and yields $Nm\nu - Nm\pi^2\nu^3/2$, the four remaining terms give $Nm\pi^2\nu^3/3$ to LO. The total result for the mass adds up to

$$M = \int dx \mathcal{H} = Nm\nu \left(1 - \frac{1}{6}\pi^2\nu^2\right), \quad (2.73)$$

in agreement with the first two terms of the Taylor expansion in $\nu$ of the exact result,

$$M = \frac{Nm}{\pi} \sin \pi\nu. \quad (2.74)$$

Since we cannot expect more from a LO calculation, the test was successful.

Now consider the question of baryon number. One interesting aspect of the NJL$_2$ model is the fact that a topologically non-trivial mean field can induce fermion number in the Dirac sea [28, 29]. Since this is a physical effect but we have eliminated the Dirac sea, it must show up somewhere else in the effective theory. To understand what is happening, we treat the fermion density $\psi^\dagger \psi = \bar{\psi}\gamma^0\psi$ as an effective operator, in analogy with the scalar or pseudoscalar densities above. To this end we write down the dressed tadpole with a $\gamma^0$ vertex and again resolve it into positive and negative energy contributions. The diagrams involving only " $-$ " terms add up to the divergent density of the (vacuum) Dirac sea and can be subtracted. The " $+$ " tadpole represents the density contribution of the valence level, $\rho_{\text{val}} = n\psi_0^\dagger\psi_0$, see Fig. 2.11a. Its first correction is driven by the two-loop graphs in Fig. 2.11. If the upper vertex is 1 (scalar " $+$ " loop, Fig. 2.11b), the vacuum polarization bubble is the time component of a four-vector and therefore proportional to $k^0$ (where $k$ is the external momentum). Since the self-energy insertion is static, this term vanishes. This is the reason why we did not have to

## 2.1. CONSTRUCTION AND VALIDATION OF NO-SEA EFFECTIVE THEORY

worry about induced fermion number in the GN$_2$ model. If the upper vertex is $i\gamma_5$ (pseudoscalar "+" loop, Fig. 2.11c), the "−" bubble behaves like the space component of a four-vector and is proportional to $k^1$. We can replace the pseudoscalar "+" bubble by the full pseudoscalar self-energy $P$, thereby summing up all the LO diagrams as explained above. The result for the induced fermion density in coordinate space then assumes the simple form $\rho_{\text{ind}} = \sqrt{N/\pi}\partial_x \Pi$, where the derivative reflects the above mentioned $k^1$-dependence. The total fermion density in the effective theory consists of the valence fermion density and the induced one,

$$\rho = \rho_{\text{val}} + \rho_{\text{ind}} = n\psi_0^\dagger \psi_0 + \sqrt{\frac{N}{\pi}}\partial_x \Pi. \tag{2.75}$$

This allows us to correctly compute the fermion density without Dirac sea. If we insert the expressions for the Shei bound state, we get an exact cancellation between the two contributions, in agreement with two independent recent results [27, 15].

Let us now switch on the bare fermion mass in the NJL$_2$ model. As discussed above, this modifies the vacuum gap equation, see Eq. (2.42), and thereby the effective couplings. For the scalar coupling the corresponding LO result can be read off from the $(\bar{\psi}\psi)^2$ term in Eq. (2.43),

$$g_{\text{eff}}^2 = \frac{\pi}{N}\frac{1}{(1+\gamma)}. \tag{2.76}$$

An analogous calculation for the pseudoscalar effective coupling yields

$$G_{\text{eff}}^2 = \frac{4m^2}{4m^2\gamma - k^2}\frac{\pi}{N}. \tag{2.77}$$

The massless pion pole gets replaced by a massive one with the pion mass

$$m_\pi^2 = 4m^2\gamma, \tag{2.78}$$

in agreement with the leading order pion mass prediction near the chiral limit (GOR relation [9, 24]). In result, the LO effective Lagrangian for the NJL$_2$ model with bare mass term becomes

$$\mathcal{L}_{\text{eff}} = \bar{\psi}(i\slashed{\partial} - m)\psi + \frac{\pi}{2N}\frac{1}{(1+\gamma)}(\bar{\psi}\psi)^2 + \frac{2\pi}{N}\bar{\psi}i\gamma_5\psi\frac{1}{\Box + m_\pi^2 - \frac{4\pi}{N(1+\gamma)}\bar{\psi}\psi}\bar{\psi}i\gamma_5\psi. \tag{2.79}$$

The next steps depend on the regime one is interested in. Two cases are particularly simple and will be considered here: i) $m_\pi^2$ and $k^2$ are both of $O(\epsilon)$, and the scalar and pseudoscalar condensates are $O(\epsilon)$ and $O(\epsilon^{3/2})$ respectively as for the Shei bound state. ii) $\gamma \gg 1$, i.e., $\pi$ and $\sigma$ masses are comparable and of $O(m)$, cf. the exact relationship [30]

$$\gamma = \frac{1}{\sqrt{\eta-1}}\arctan\frac{1}{\sqrt{\eta-1}}, \quad \eta = \frac{4m^2}{m_\pi^2}. \tag{2.80}$$

In case i) we can neglect $\gamma$ in the scalar effective coupling multiplying $(\bar{\psi}\psi)^2$ since it is of higher order than $O(\epsilon^2)$. The same diagrams contribute to the pseudoscalar terms as before, except that the pion propagator becomes massive everywhere, $\Box \to \Box + m_\pi^2$. The cancellation in the $\pi\pi$ scattering amplitude illustrated in Fig. 2.9 is upset by the bare mass term, as expected from chiral perturbation theory and low energy theorems. However, since the correction gets an additional factor of $\gamma$, these terms can again be ignored to LO. As a result, the non-local effective action (2.62) of the massless NJL$_2$ model now gets replaced by

$$\mathcal{L}_{\text{eff}} = \bar{\psi}(i\slashed{\partial} - m)\psi + \frac{\pi}{2N}(\bar{\psi}\psi)^2 + \frac{2\pi m^2}{N}\bar{\psi}i\gamma_5\psi \frac{1}{\Box + m_\pi^2 - \frac{4\pi m}{N}\bar{\psi}\psi}\bar{\psi}i\gamma_5\psi, \quad (2.81)$$

with $m_\pi^2$ defined in Eq. (2.78). This can again be converted into a local effective action by introducing an elementary pseudoscalar field $\Pi$. The only modification as compared to Eq. (2.63) is the appearance of a mass term for $\Pi$,

$$\mathcal{L}_{\text{eff}}^{\text{loc}} = \bar{\psi}(i\slashed{\partial} - m)\psi + \frac{\pi}{2N}(\bar{\psi}\psi)^2 + \frac{1}{2}\partial_\mu\Pi\partial^\mu\Pi + \sqrt{\frac{4\pi}{N}}m\bar{\psi}i\gamma_5\psi\Pi + \frac{2\pi m}{N}\bar{\psi}\psi\Pi^2 - \frac{1}{2}m_\pi^2\Pi^2. \quad (2.82)$$

This opens up the possibility to extend Shei's bound state to finite bare fermion masses. Of course one would then have to verify that our assumptions concerning the order of $\epsilon$ of the valence condensates are fulfilled.

In case ii) we go into a regime where both $\pi$ and $\sigma$ propagators may be treated as point-like. The scalar and pseudoscalar effective couplings become equal (to LO),

$$g_{\text{eff}}^2 = G_{\text{eff}}^2 = \frac{\pi}{N\gamma}, \quad \gamma \gg 1. \quad (2.83)$$

All higher order corrections are suppressed by inverse powers of $\gamma$, and the LO effective Lagrangian assumes the same structure as the original Lagrangian (apart from the mass term and the replacement of $g^2$ by the effective coupling),

$$\mathcal{L}_{\text{eff}} = \bar{\psi}\left(i\slashed{\partial} - m\right)\psi + \frac{\pi}{2N\gamma}\left[(\bar{\psi}\psi)^2 + (\bar{\psi}i\gamma_5\psi)^2\right]. \quad (2.84)$$

In this regime, the LO effective Lagrangian of the GN$_2$ model reduces to

$$\mathcal{L}_{\text{eff}} = \bar{\psi}\left(i\slashed{\partial} - m\right)\psi + \frac{\pi}{2N\gamma}(\bar{\psi}\psi)^2. \quad (2.85)$$

These last two equations may be considered as the extreme heavy-fermion (or non-relativistic) limit of the original field theories.

Finally, we comment on some early work which is relevant in this context. In the 70's, there was some debate about the use of "classical calculations" for fermions in field theory. These were mean-field calculations without Dirac sea, solving non-linear Dirac-HF equations [16, 17]

## 2.1. CONSTRUCTION AND VALIDATION OF NO-SEA EFFECTIVE THEORY

exactly as in our "no-sea" effective theory. What was done at that time simply amounted to use the original Lagrangian with a mass term added and replacing the bare coupling constant by an effective one. Shei [13] noted that this type of calculation gave reasonable results for the $GN_2$ model, but was completely off in the case of the $NJL_2$ model. We can now understand the reasons behind this observation, using the framework of effective field theory. In the $GN_2$ model, the lowest order effective Lagrangian happens to agree with the ad-hoc prescription used in Refs. [16, 17]. This explains why it gave useful results, at least for small filling fraction. Incidentally, another example of a "no-sea" calculation for this model can be found in Ref. [14] where the effective coupling constant $g_{\text{eff}}^2 = \pi/N$ was determined "phenomenologically", see Eq. (4.17) of that paper. We have now derived this number microscopically and are able to improve the calculation in a systematic fashion. In the vicinity of the chiral limit of the $NJL_2$ model, the naive recipe for writing down the effective action breaks down due to the pion pole. One cannot ignore the singular momentum dependence of the effective pseudoscalar coupling ($\sim 1/k^2$) which gives rise to long range non-localities. This is the reason for the observed discrepancy between short range "no-sea" and long range semi-classical mean fields. However, in the heavy-fermion limit, we do recover an effective action which agrees with the simple recipe of trading the bare fermion mass for the physical fermion mass in the vacuum and introducing an effective coupling, see Eq. (2.84). Hence the technically rather nice classical calculations of the $NJL_2$ model done in the 70's (which had to fail in the chiral limit, as we now understand) can be invoked to solve the $NJL_2$ model in the heavy-fermion limit $\gamma \gg 1$. Here, the no-sea effective theory Lagrangian, Eq. (2.85), exactly resembles that of the original theory, Eq. (2.1) with $\lambda = 1$. In the representation of the $\gamma$-matrices (2.37), the bound state found by Lee et al. [17] has the spinor wave function (filling fraction $\nu$, energy $E_0$)

$$\psi_0 = \sqrt{\frac{(m - E_0)}{2G}} \frac{1}{\cosh^2 \xi + \alpha^2 \sinh^2 \xi} \begin{pmatrix} \alpha \sinh \xi + \cosh \xi \\ \alpha \sinh \xi - \cosh \xi \end{pmatrix},$$

$$G = \frac{\pi \nu}{2\gamma}, \quad \alpha = \tan \frac{G}{2},$$

$$E_0 = m \cos G, \quad \xi = mx \sin G. \tag{2.86}$$

We have checked that it satisfies the (classical) non-linear Dirac equation obtained from the effective Lagrangian (2.84) exactly. The baryon mass is given by

$$M_B = nm \frac{\sin G}{G} = nm \left( 1 - \frac{\pi^2 \nu^2}{24\gamma^2} + \cdots \right) \tag{2.87}$$

Due to the truncation of the effective action, only the two listed terms of the series expansion can be trusted. At the same time the induced fermion density gets suppressed by a factor of $1/\gamma$ as compared to the valence density. The corresponding result for the $GN_2$ model with effective action (2.85) can be read off Eq. (2.47) (for $\gamma \gg 1$) and agrees with (2.87) to this order in $\nu$. As a matter of fact, the spinors $\psi_0$ also become equal in the limit $\gamma \to \infty$, as can be seen by expanding in powers of $\nu$. The reason can be traced back to the fact that the ratio of pseudoscalar to scalar condensates decreases like $1/\gamma$. Thus we learn that in the heavy-fermion limit, the difference between discrete and continuous chiral symmetry of the four-fermion interaction becomes less and less important, and baryons of the $GN_2$ and $NJL_2$ model approach each other.

What makes the $NJL_2$ model with bare mass especially interesting is the fact that, in contrast to the $GN_2$ model (with and without $m_0$) as well as the $NJL_2$ model without bare mass term, no exact analytical solutions are known so far. Let us therefore concentrate on this model and try to gain some further insight by means of our effective theory. Once again, we focus on baryons.

## 2.2 Application of no-sea effective theory - new results

Recall that the results which could be obtained from effective theory calculations in the preceding section, were valid up to a certain order in a given expansion parameter only. Having specified an expansion parameter, one has to ensure that all effective interaction terms are consistently incorporated up to a desired order in this parameter. To increase the order of validity, one obviously has to include more and more interaction terms in the effective Lagrangian.

As no exact solution of the $NJL_2$ model with bare mass is known, there is no a-priori measure which allows us to systematically classify and organize the different contributions. We proceed as follows. As far as terms involving only scalar interactions are concerned, we keep all contributions in Eq. (2.43). The terms involving pseudoscalar interactions require one additional approximation. Starting point is the LO effective Lagrangian for the massive $NJL_2$ model, Eq. (2.79). If we insist on an analytical solution as we do in this section, we are only able to handle the case in which the non-locality due to one-pion exchange is of short range. We therefore assume that the mass term $m_\pi^2 = 4\gamma$ dominates the inverse pion propagator and expand in the remaining two terms as follows,

$$\frac{2\pi}{N} \bar{\psi} i\gamma_5 \psi \frac{1}{\Box + m_\pi^2 - \frac{4\pi}{N(1+\gamma)} \bar{\psi}\psi} \bar{\psi} i\gamma_5 \psi$$
$$\approx \frac{2\pi}{N} \bar{\psi} i\gamma_5 \psi \frac{1}{m_\pi^2} \left(1 - \frac{\Box}{m_\pi^2} + \frac{4\pi \bar{\psi}\psi}{N m_\pi^2 (1+\gamma)}\right) \bar{\psi} i\gamma_5 \psi. \qquad (2.88)$$

The result is added to the GN model effective Lagrangian. At this stage it is difficult to predict the precise range of applicability of the approximation due to the unavoidable issue of self-consistency. We will return to this question once we have solved the non-linear Dirac-HF

## 2.2. APPLICATION OF NO-SEA EFFECTIVE THEORY - NEW RESULTS

equation. The final effective Lagrangian reads ($m = 1$)

$$\mathcal{L}_{\text{eff}} = \bar{\psi}(i\partial\!\!\!/ - 1)\psi + \frac{\pi}{2N}\frac{1}{(1+\gamma)}(\bar{\psi}\psi)^2 - \frac{\pi}{24N}\frac{1}{(1+\gamma)^2}(\Box\bar{\psi}\psi)(\bar{\psi}\psi)$$
$$+ \frac{\pi^2}{6N^2}\frac{1}{(1+\gamma)^3}(\bar{\psi}\psi)^3 + \frac{11\pi}{1440N}\frac{(1+6\gamma/11)}{(1+\gamma)^3}(\Box^2\bar{\psi}\psi)(\bar{\psi}\psi)$$
$$- \frac{\pi^2}{12N^2}\frac{(1+\gamma/2)}{(1+\gamma)^4}(\Box\bar{\psi}\psi)(\bar{\psi}\psi)^2 + \frac{\pi^3}{6N^3}\frac{(1+\gamma/4)}{(1+\gamma)^5}(\bar{\psi}\psi)^4 + \frac{\pi}{2N\gamma}(\bar{\psi}i\gamma_5\psi)^2$$
$$- \frac{\pi}{8N\gamma^2}(\Box\bar{\psi}i\gamma_5\psi)(\bar{\psi}i\gamma_5\psi) + \frac{\pi^2}{2N^2\gamma^2}\frac{1}{(1+\gamma)}(\bar{\psi}i\gamma_5\psi)^2(\bar{\psi}\psi). \quad (2.89)$$

By construction, it is to be used in the positive energy sector only, since the effects of the negative energy states are already encoded in the Lagrangian. For a single baryon, the ensuing Euler-Lagrange equation is the non-linear Dirac equation for the (normalized) valence spinor,

$$\left(-\gamma_5 i\partial_x + \gamma^0 S + i\gamma^1 P\right)\psi_0 = E_0\psi_0, \quad \int dx\, \psi_0^\dagger \psi_0 = 1. \quad (2.90)$$

Assuming the valence level to be occupied with filling fraction $\nu = n/N$ (where $n$ is the valence fermion number) and using $\eta = \pi\nu$ for ease of notation (all $\pi$'s disappear), the scalar and pseudoscalar potentials are expressed self-consistently as ($' = \partial_x$)

$$S = 1 - \frac{\eta s_0}{(1+\gamma)} - \frac{\eta s_0''}{12(1+\gamma)^2} - \frac{\eta^2 s_0^2}{2(1+\gamma)^3} - \frac{11\eta(1+6\gamma/11)s_0^{IV}}{720(1+\gamma)^3}$$
$$- \frac{\eta^2(1+\gamma/2)}{6(1+\gamma)^4}\left(2s_0'' s_0 + (s_0')^2\right) - \frac{2\eta^3(1+\gamma/4)s_0^3}{3(1+\gamma)^5} - \frac{\eta^2 p_0^2}{2\gamma^2(1+\gamma)},$$
$$P = -\frac{\eta p_0}{\gamma} - \frac{\eta p_0''}{4\gamma^2} - \frac{\eta^2 p_0 s_0}{\gamma^2(1+\gamma)} \quad (2.91)$$

through scalar ($s_0$) and pseudoscalar ($p_0$) valence level condensates

$$s_0 = \bar{\psi}_0\psi_0, \quad p_0 = \bar{\psi}_0 i\gamma_5 \psi_0. \quad (2.92)$$

It seems hopeless to solve the complicated non-linear Dirac equation (2.90)-(2.92) in closed analytical form. However, since the Lagrangian (2.89) is an approximate one, it is sufficient to solve the equation perturbatively in some formal expansion parameter. We choose to expand in $\nu$ and postpone till later the discussion of the validity of the truncation scheme. In order to solve the Dirac equation, we then find that we first have to solve a simple non-linear differential equation, followed by a sequence of inhomogeneous, linear differential equations. The lengthy analytical computations can be done conveniently with computer algebra (we used MAPLE), therefore we skip the details and record only the final results. The baryon mass (computed from the classical field energy) has the following power series expansion in $\eta = \pi\nu$,

$$\pi\frac{M_B}{N} = \eta - \frac{\eta^3}{24(1+\gamma)^2} + \frac{(\gamma^2 - 15\gamma - 8)\eta^5}{1920\gamma(1+\gamma)^5} - \frac{(\gamma^4 - 302\gamma^3 + 265\gamma^2 + 376\gamma + 88)\eta^7}{322560\gamma^2(1+\gamma)^8}. \quad (2.93)$$

Here we kept terms up to $\mathcal{O}(\eta^7)$. The energy $E_0$ of the valence level is closely related to $M_{\rm B}$. According to standard HF theory, the single particle energy can be interpreted as removal energy of a fermion, or equivalently

$$E_0 = \frac{\partial M_{\rm B}}{\partial n}. \tag{2.94}$$

This is indeed what we find in the calculation, so that there is no need to spell out $E_0$. Next we turn to the HF potentials (2.91) derived from the self-consistent valence level spinor $\psi_0$. For the scalar and pseudoscalar potentials, we obtain

$$\begin{aligned} S &= 1 + \frac{s_{22}}{\cosh^2 \xi}\eta^2 + \left(\frac{s_{42}}{\cosh^2 \xi} + \frac{s_{44}}{\cosh^4 \xi}\right)\eta^4 + \left(\frac{s_{62}}{\cosh^2 \xi} + \frac{s_{64}}{\cosh^4 \xi} + \frac{s_{66}}{\cosh^6 \xi}\right)\eta^6 + \cdots \\ P &= \left[\frac{p_{33}}{\cosh^3 \xi}\eta^3 + \left(\frac{p_{53}}{\cosh^3 \xi} + \frac{p_{55}}{\cosh^5 \xi}\right)\eta^5 + \cdots\right]\sinh \xi \end{aligned} \tag{2.95}$$

with $\gamma$-dependent coefficients $s_{mn}, p_{mn}$ relegated to appendix A.1 (the subscripts $n, m$ denote the powers of $\eta$ and $1/\cosh \xi$, respectively). The spatial variable $\xi = yx$ in Eq. (2.95) involves a scale factor

$$y = \frac{\eta}{2(1+\gamma)} - \frac{(\gamma^2 - 3\gamma - 2)\eta^3}{48\gamma(1+\gamma)^4} + \frac{(3\gamma^4 - 226\gamma^3 - 65\gamma^2 + 68\gamma + 24)\eta^5}{11520\gamma^2(1+\gamma)^7} + \cdots. \tag{2.96}$$

One verifies that it is related to the single particle energy $E_0$ via

$$E_0 = \sqrt{1 - y^2}. \tag{2.97}$$

This last relation can be understood in physics terms as follows. Asymptotically, the valence wave function falls off exponentially with the $\kappa$ value (imaginary momentum) of the bound state. Due to self-consistency, the same parameter will govern the asymptotic exponential decay of the potentials in a no-sea HF calculation. This is exactly what Eq. (2.97), together with the shape of the potentials $S$ and $P$, guarantees.

We have tested the above results in two different ways. If we switch off the pseudoscalar coupling, we can carry out the same calculation for the massive GN model and reproduce the well-known exact results to the expected accuracy. A more specific test of the NJL$_2$ model calculation is provided by the divergence of the axial current, Eq. (2.5), implying

$$\partial_x \rho = -\frac{2\gamma}{\pi} P. \tag{2.98}$$

Here $\rho$ denotes the fermion density per flavor consisting of the valence density and the induced part from the Dirac sea, cf. Eq. (2.75),

$$\rho = \nu \psi_0^\dagger \psi_0 - \frac{1}{2\pi}\partial_x P. \tag{2.99}$$

## 2.2. APPLICATION OF NO-SEA EFFECTIVE THEORY - NEW RESULTS

Eqs. (2.98) and (2.99) yield the following non-trivial identity relating the valence fermion density to the pseudoscalar HF potential,

$$\nu \partial_x (\psi_0^\dagger \psi_0) = -\frac{2\gamma}{\pi} P + \frac{1}{2\pi} \partial_x^2 P . \tag{2.100}$$

In our calculation it is violated at order $\nu^7$. For the sake of completeness, we quote the separate results for the valence fermion density $\rho_{\text{val}} = \nu \psi_0^\dagger \psi_0$,

$$\pi \rho_{\text{val}} = \frac{v_{22}}{\cosh^2 \xi} \eta^2 + \left( \frac{v_{42}}{\cosh^2 \xi} + \frac{v_{44}}{\cosh^4 \xi} \right) \eta^4$$
$$+ \left( \frac{v_{62}}{\cosh^2 \xi} + \frac{v_{64}}{\cosh^4 \xi} + \frac{v_{66}}{\cosh^6 \xi} \right) \eta^6 + \cdots , \tag{2.101}$$

and for the induced fermion density $\rho_{\text{ind}} = -\frac{1}{2\pi} \partial_x P$,

$$\pi \rho_{\text{ind}} = \left( \frac{i_{42}}{\cosh^2 \xi} + \frac{i_{44}}{\cosh^4 \xi} \right) \eta^4 + \left( \frac{i_{62}}{\cosh^2 \xi} + \frac{i_{64}}{\cosh^4 \xi} + \frac{i_{66}}{\cosh^6 \xi} \right) \eta^6 + \cdots . \tag{2.102}$$

The coefficients can again be found in appendix A.1.

Let us now discuss the range of validity of our truncation. The most obvious candidate is the regime

$$\gamma \gg 1 \quad \text{(regime I)} \tag{2.103}$$

and arbitrary $\nu$. In this case we can read off from our results that $\bar{\psi}\psi \sim 1/\gamma$, $\bar{\psi} i\gamma_5 \psi \sim 1/\gamma^2$ and $\Box \sim 1/\gamma^2$. Inspection of the coefficients in $\mathcal{L}_{\text{eff}}$ then shows that we have kept all terms up to order $1/\gamma^8$ in an asymptotic expansion for large $\gamma$. The identity (2.100) is violated at order $1/\gamma^6$. This is due to the fact that we can trust the result for $P$ (expanded in $1/\gamma$) up to order $1/\gamma^6$ only. A careful examination of our calculation unveils that the terms written explicitly in Eq. (2.102) for $\rho_{\text{val}}$ are correct up to order $1/\gamma^6$. By way of example, we note the asymptotic behavior of $M_B$ for $\gamma \to \infty$ and full occupation of the valence level ($\nu = 1$) complementary to Eq. (9) of [31],

$$\frac{M_B}{N} = 1 - \frac{\pi^2}{24\gamma^2} + \frac{\pi^2}{12\gamma^3} - \frac{\pi^2(240 - \pi^2)}{1920\gamma^4}$$
$$+ \frac{\pi^2(16 - \pi^2)}{96\gamma^5} - \frac{\pi^2(67200 - 13776\pi^2 + \pi^4)}{322560\gamma^6} . \tag{2.104}$$

Another case of interest is weak occupation of the valence level, i.e., taking the formal expansion in $\nu$ literally. Here, $\bar{\psi}\psi \sim \nu^2$, $\bar{\psi} i\gamma_5 \psi \sim \nu^3$, $\Box \sim \nu^2$ and all terms up to order $\nu^8$ are kept in $\mathcal{L}_{\text{eff}}$. A necessary condition for this to be valid is $\nu \ll 1$. Due to the expansion of the propagator in Eq. (2.88), inverse powers of $\gamma$ appear in the coefficients, and we have to require in addition $\nu^2 \ll \gamma$. In this regime, all terms are then consistently kept and $M_B$ is valid up to $\mathcal{O}(\nu^7)$. The validity of Eq. (2.100) is again limited by $P$, which is trustworthy up to order $\nu^6$ only. A

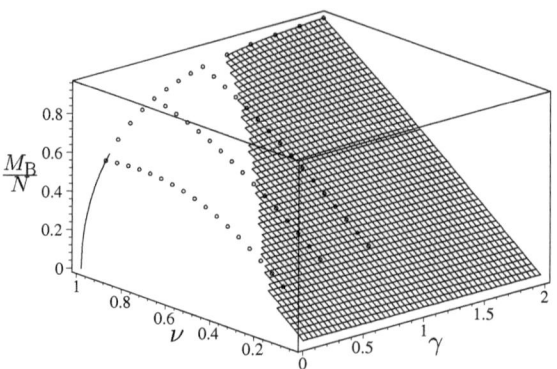

Figure 2.12: Baryon mass as a function of filling fraction $\nu$ and confinement parameter $\gamma$. Numerical HF results (circles) together with analytical results: Shown are only results which are stable at the 0.2% level if one removes the highest order correction term. Curve starting at $\gamma = 0, \nu = 1$: derivative expansion. Tilted surface: No-sea effective theory.

careful examination of our calculation assures that we can trust all the terms kept explicitly in Eq. (2.102) for $\rho_{\text{val}}$. This 2nd regime of applicability may be summarized as

$$\nu \ll \min(1, \sqrt{\gamma}) \qquad \text{(regime II)}. \qquad (2.105)$$

If the condition $\nu \ll \sqrt{\gamma}$ does not hold, the non-locality of the effective action becomes long range and our approximation breaks down.

We conclude this section by placing our findings in the context. Recently, much effort has been invested to establish the phase diagram of the massive NJL$_2$ model [32, 33]. In this task the baryons form an important building block. Before invoking the no-sea effective theory, only the derivative expansion provided some analytical insight [22, 31]. However, its range of applicability is limited to $\nu = 1$ and small $\gamma$ only.

From a practical point of view, it is fairly easy to judge the quality of an effective theory by comparing successive orders in the expansion. This is illustrated in Fig. 2.12 where we plot the baryon mass (per flavor) $M_{\text{B}}/N$ as computed from Eqs. (9) of [31] and (2.93) versus $\nu$ and $\gamma$. We only display the result if the highest order correction kept contributes less than 0.002 of the total result (other numerical values for the tolerance would give qualitatively similar plots). Needless to say, the no-sea calculation is also valid at higher values of $\gamma$ not shown here. Note that the structure of the baryons is very different, depending on which asymptotic expansion we look at. Baryons described by the derivative expansion have no filled positive energy level, so that baryon number is entirely due to fermion number induced in the Dirac

## 2.3. CONCLUSIONS AND OUTLOOK

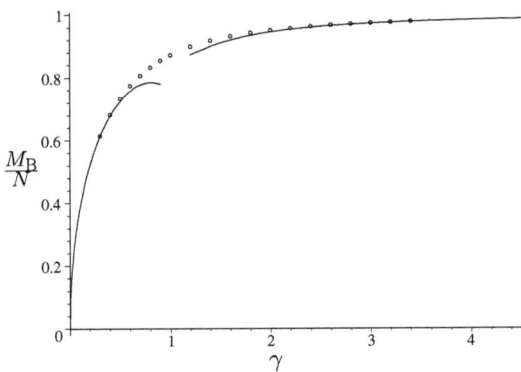

Figure 2.13: Circles: numerical results for baryon mass at $\nu = 1$ versus $\gamma$. Curves: analytical asymptotic predictions (derivative expansion to the left, no-sea effective theory to the right).

sea. If on the other hand the no-sea effective theory applies, there has to be a positive energy valence level, and the valence fermion density dominates over the induced one, cf. the powers of $\eta$ in Eqs. (2.101) and (2.102). Our formulae also show that in the limit of large $\gamma$ for fixed $\nu$, or in the limit of small $\nu$ for fixed $\gamma$, the pseudoscalar terms are suppressed and the NJL$_2$ results converge to those of the GN model. In fact, a non-relativistic approximation to the Dirac equation would be perfectly adequate in this regime. The region where the uppermost filled level reaches the middle of the gap and crosses zero energy lies in the white part of Fig. 2.12. Here only a numerical treatment is possible so far.

Regarding the phase diagram, the question of most immediate interest is the dependence of $M_B$ on $\gamma$ for full occupation $\nu = 1$. This is expected to yield a critical (2nd order) curve in the phase diagram at $T = 0$. The results in this particular region are depicted in Fig. 2.13.

## 2.3 Conclusions and outlook

Using exactly solvable four-fermion models as theoretical laboratory, we have demonstrated that "integrating out the Dirac sea" is indeed a viable concept. Tractable effective Lagrangians could be derived, valid for multi-fermion bound states with weakly occupied valence level or large bare fermion masses. Whereas the original semi-classical calculations of bound states in GN$_2$ and NJL$_2$ models required sophisticated and highly specialized techniques to deal with the Dirac sea, the effective theories amount to solving the classical Euler-Lagrange equation, here a non-linear Dirac equation. One might suspect that the difficulties of the full non-perturbative calculations come back in the derivation of the effective Lagrangian, but this is not quite so. Although this derivation has non-perturbative features in the form of unavoidable resummations, the actual calculation is based on standard one-loop Feynman diagrams and thus perturbative. This suggests that similar techniques can also be applied to more realistic, higher dimensional theories.

It seems to be a natural step to switch to the GN model family in higher dimensions first. However, turning into this direction, one gets confronted with several difficulties. In 3+1 dimensions the GN model family is no longer renormalizable, while there is no chiral symmetry in 2+1 dimensions. Turning from the $GN_2$ to the $GN_3$ model, the Lagrangian can be retained and one can principally work with a two dimensional representation of the Dirac algebra, even if there is no $\gamma^5$-matrix. One should note that in this way, the $GN_3$ model can be treated in the very same way as $GN_2$. This opens up the possibility to search for baryons in a no-sea effective theory of the $GN_3$ model. However, $GN_3$ is usually formulated by introducing additional isospin degrees of freedom. Obviously, the $NJL_3$ model has to be defined in a completely different way as compared to $NJL_2$ [34]. Rather then going into these peculiarities in 2+1 dimensions, in the next chapter we turn directly to a different, physically more realistic and broadly studied field theory, namely the Walecka model. It is renormalizable also in 3+1 dimensions and yields interesting results within the HF approximation.

To conclude this chapter, let us summarize our findings obtained in the course of "integrating out the Dirac sea" in the GN model family.

In the $GN_2$ case, we were able to derive the NNLO approximation in some small expansion parameter. Since the exact DHN baryons and their generalization to finite bare fermion mass are known, we could show that the resulting effective theory works quantitatively. Thus baryon binding energies can be predicted "classically" in the full range of filling fraction and bare mass at the 1% level, quite unexpectedly for us. As far as the formalism is concerned, the main lesson we learned is that the bare parameters of the original theory $(m_0, g^2)$ disappear owing to resummations. It would not make sense to truncate the procedure at any fixed number of loops. All questions of renormalization and UV divergences are then properly dealt with and do not show up any more in the effective theory, which is of course finite.

In the $NJL_2$ case, when applying the same strategy we ran into a new difficulty. The pseudoscalar effective coupling develops a singularity at $k^2 = 0$ due to the massless pion pole. Although one can again identify the LO of a systematic expansion in a small parameter, this necessarily causes a long range non-locality of the effective Lagrangian. The purely fermionic, non-local effective theory can be cast into a more convenient "almost local" form by introducing an additional elementary pion field. The resulting theory contains elementary fermions and pions but is fully consistent, judging from the derivation and the comparison with Shei's bound

## 2.3. CONCLUSIONS AND OUTLOOK

states. The role of the massless pion explains the conspicuous difference between short range mean fields in the $GN_2$ model and long range mean fields in the $NJL_2$ model, which had caused headaches in early, more naive attempts to treat fermions classically. Moreover, we could show how the important phenomenon of "induced fermion number" can be recovered in a "no-sea" effective theory.

Furthermore, as an application of our no-sea effective theory, some completely new analytical insight concerning baryons could be obtained in the massive $NJL_2$ model. As noted, these findings are also relevant for the $NJL_2$ phase diagram.

Finally, notice that no attempt was made to reduce the Dirac equation to a non-relativistic Schrödinger equation. Although this would be possible, we feel that it would only make our formulae more messy. Our main focus here is not on relativistic kinematics, but on the dynamical effects of the Dirac sea.

# Chapter 3

# The Walecka model

The relativistic approach to nuclear matter was pioneered by Walecka [1], who formulated a simple relativistic quantum field theory in 3+1 dimensions, mimicking the main features of the nucleon-nucleon interaction: Forces between nucleons (spinor field $\psi$) are mediated by the exchange of mesons. As far as isospin symmetric matter (equal number of protons and neutrons) is concerned, the dominant one meson exchanges come from the scalar sigma ($\sigma$) meson and vector omega ($\omega$) meson. The corresponding model subsequently came to be known as Walecka model, ($\sigma$, $\omega$) model or Quantum Hadrodynamics I (QHD-I). It is defined by the following Lagrangian [1, 35]

$$\mathcal{L} = \bar{\psi}\big(\gamma_\mu(\mathrm{i}\partial^\mu - \tilde{g}_\omega \omega^\mu) - (\tilde{m} - \tilde{g}_\sigma \sigma)\big)\psi + \frac{1}{2}\big(\partial_\mu \sigma \partial^\mu \sigma - \tilde{m}_\sigma^2 \sigma^2\big)$$
$$+ \frac{1}{3!}\tilde{c}_3 \sigma^3 + \frac{1}{4!}\tilde{c}_4 \sigma^4 - \frac{1}{4}F_{\mu\nu}F^{\mu\nu} + \frac{1}{2}\tilde{m}_\omega^2 \omega_\mu \omega^\mu, \quad (3.1)$$

where $F^{\mu\nu} = \partial^\mu \omega^\nu - \partial^\nu \omega^\mu$ is the kinetic term associated with the vector field $\omega^\mu$. We assume $N$ fermion flavors. The "tilded" quantities denote the bare parameters of the theory. Note that we will not allow for physical scalar self-interactions $\sim \sigma^3$ and $\sim \sigma^4$ in the following. However, the cubic and quartic scalar self-interactions have to be included in the Lagrangian in 3+1 dimensions, to make the model renormalizable.

In order to gain any analytical insights into the Walecka model, one has to specify an approximation scheme. As the physical coupling constants $g_\sigma$ and $g_\omega$ are expected to be large, a perturbative approach would be inadequate. However, mean-field theory (MFT), the relativistic Hartree approximation (RHA) and the Hartree-Fock (HF) approximation are known to yield physically sensible results [36]. This makes the Walecka model an ideal testing ground for our no-sea effective theory approach. Here we base our approach on the RHA, which coincides with the approximation used in the previous chapter.

The Walecka model was originally formulated in 3+1 dimensions [1]. However, there are also studies in 1+1 dimensions, where calculational efforts are less and even some exact solutions

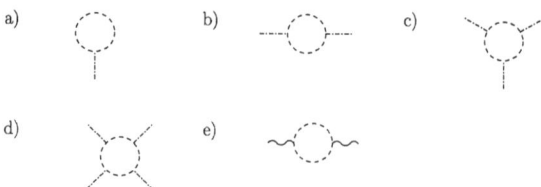

Figure 3.1: Divergent primitive Feynman diagrams. In 3+1 dimensions, diagrams a)-e) diverge. In 1+1 dimensions, diagrams a), b) and e) diverge, while c) and d) are finite. Dashed line: free fermion propagator; dash-dotted line: scalar propagator; curly line: vector propagator.

are known [37]. As can be seen by naive dimensional analysis already, in calculating physical observables there will be UV divergences. While $[g_\sigma] = [g_\omega] = M^0$ in 3+1 dimensions, in 1+1 dimensions $[g_\sigma] = [g_\omega] = M^1$. This indicates that the theory is renormalizable in 3+1 and super-renormalizable in 1+1 dimensions. Some renormalization procedure is necessary in both cases. The divergent primitive Feynman diagrams are depicted in Fig. 3.1. While all diagrams a)-e) diverge in 3+1 dimensions, only a), b) and e) diverge in 1+1 dimensions. Note that the fermion tadpole diagram coupled via a vector propagator vanishes in the vacuum.

In the previous chapter about the GN model family, all divergences could be eliminated through the fermion mass gap equation in the vacuum. Due to the presence of the dynamic meson fields, this is no longer the case here. The Walecka model is conventionally dealt with using counter term renormalization, a standard technique in the renormalization of interacting QFTs. To exhibit the relationship with the method used in the previous chapter, we show how to treat the GN model with counter term renormalization in appendix A.2. Let us now focus on the Walecka model. The original Lagrangian (3.1) is written as $\mathcal{L} = \mathcal{L}' + \mathcal{L}_{CT}$, with

$$\mathcal{L}' = \bar{\psi}\big(\gamma_\mu(i\partial^\mu - g_\omega\omega^\mu) - (m - g_\sigma\sigma)\big)\psi + \frac{1}{2}\left(\partial_\mu\sigma\partial^\mu\sigma - m_\sigma^2\sigma^2\right)$$
$$+ \frac{1}{3!}c_3\sigma^3 + \frac{1}{4!}\tilde{c}_4\sigma^4 - \frac{1}{4}F_{\mu\nu}F^{\mu\nu} + \frac{1}{2}m_\omega^2\omega_\mu\omega^\mu \qquad (3.2)$$

and the counter term Lagrangian

$$\mathcal{L}_{CT} = -\delta_m\bar{\psi}\psi + \frac{1}{2!}\alpha_2\sigma^2 + \frac{1}{3!}\alpha_3\sigma^3 + \frac{1}{4!}\alpha_4\sigma^4 + \frac{1}{2}\alpha_\sigma\partial_\mu\sigma\partial^\mu\sigma + \frac{1}{4}\alpha_\omega F_{\mu\nu}F^{\mu\nu}. \qquad (3.3)$$

While the tilded parameters in Eq. (3.1) are bare and unphysical, in Eq. (3.2) the analogous (untilded) parameters are considered as physical and measurable. In this sense, the counter term Lagrangian is not added. The original Lagrangian is rather split into two parts. $\mathcal{L}'$ is the physical part, $\mathcal{L}_{CT}$ contains the infinite but unobservable shifts between the bare and the physical parameters. The corresponding derivation can be found in appendix B.1. As their choice is scale dependent, the physical parameters have to be specified by renormalization conditions, stating under which circumstances they have to be measured. Subsequently, the counter terms are tuned such that theory and measurement agree under these conditions.

Note that the individual counter terms can be unambiguously assigned to the divergent primitive diagrams depicted in Fig. 3.1. Choosing the counter terms so as to cancel the divergences, the theory is rendered finite. In result, a divergent primitive Feynman diagrams never occurs on its own, but always in its renormalized, finite form.

In 1+1 dimensions, the diagrams depicted in Fig. 3.1c and Fig. 3.1d are finite. Hence, the counter term contributions $\sim \sigma^3$ and $\sim \sigma^4$ are not needed. Working in 1+1 dimensions, we therefore set $\tilde{c}_3 = \tilde{c}_4 = 0$ from the beginning. Consequently, the respective counter term contributions are not generated and $\alpha_3 = \alpha_4 = 0$.

As in the previous chapter, we want to derive a purely fermionic no-sea effective theory featuring positive energy states only. Our main aim is to proceed to the 3+1 dimensional model. However, as the transition from 1+1 to 3+1 dimensions will turn out to be rather easy and, on the other hand, fully analytical calculations are feasible in 1+1 dimensions, we find it worthwhile to consider the 1+1 dimensional Walecka model first.

## 3.1 Walecka model in 1+1 dimensions

### 3.1.1 Construction of no-sea effective theory

In 1+1 dimensions Eq. (3.1) ($\tilde{c}_3 = \tilde{c}_4 = 0$) is quadratic in the scalar field $\sigma$ as well as the vector field $\omega^\mu$. Hence, both fields can be integrated out exactly, yielding a purely fermionic theory

$$\mathcal{L} = \bar{\psi}\left(\mathrm{i}\slashed{\partial} - \tilde{m}\right)\psi + \frac{\tilde{g}_\sigma^2}{2}\bar{\psi}\psi\frac{1}{\Box + \tilde{m}_\sigma^2}\bar{\psi}\psi + \frac{\tilde{g}_\omega^2}{2}\bar{\psi}\gamma^\nu\psi\frac{1}{\Box + \tilde{m}_\omega^2}\left(-\delta_\nu^{\ \mu} - \frac{\partial_\nu\partial^\mu}{\tilde{m}_\omega^2}\right)\bar{\psi}\gamma_\mu\psi. \quad (3.4)$$

Here, the similarity to the GN model family becomes obvious. While the models share a scalar four-fermion interaction, instead of the pseudoscalar interaction in the NJL model, there is an additional vector interaction in the Walecka model. A more significant difference between the two models lies in the explicit form of the respective four-fermion interactions. While the GN model family exhibits point-like fermion-fermion interactions only, in the Walecka model, the interactions are of finite range, mediated by dynamical meson fields. The Feynman propagators for the $\sigma$ and $\omega$ fields can easily be read off Eq. (3.4). Turning to the local limit of the scalar meson propagator and setting $\tilde{g}_\omega = 0$, the GN model with $g^2 = \frac{\tilde{g}_\sigma^2}{\tilde{m}_\sigma^2}$ is recovered [2].

As in the previous chapter, let us first determine the physical fermion mass in the vacuum. In Hartree approximation, it arises as depicted in Fig. 3.2. The vector self-energy contribution vanishes in the vacuum (cf. appendix B.2). Hence, the generation of the self-consistent mass

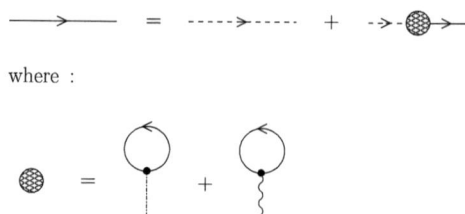

Figure 3.2: Dyson equation for fermion propagator in RHA. Dashed line: free propagator; solid line: dressed propagator; shaded circle: self-energy; dash-dotted line: scalar propagator; curly line: vector propagator.

in the vacuum closely resembles that in the GN model family.

While in the GN model, we just had the coupling constant $g^2$, associated with the point-like four-fermion interaction, in the context of the Walecka model $g^2 = \frac{\tilde{g}_\sigma^2}{\tilde{m}_\sigma^2}$ is the ratio of two distinct bare parameters in the original Lagrangian (3.1), which, after renormalization, can be associated with physical measurable parameters. As far as the mass is concerned, only the ratio enters. However, considering subleading correction terms in the scalar interaction of Eq. (3.4) to account for finite range effects,

$$\frac{\tilde{g}_\sigma^2}{2} \bar{\psi}\psi \frac{1}{\Box + \tilde{m}_\sigma^2} \bar{\psi}\psi = \frac{1}{2} \frac{\tilde{g}_\sigma^2}{\tilde{m}_\sigma^2} \bar{\psi}\psi \left(1 - \frac{\Box}{\tilde{m}_\sigma^2} + \cdots \right) \bar{\psi}\psi, \qquad (3.5)$$

$\tilde{m}_\sigma^2$ enters the calculation on its own. In contrast to the GN model family where we had two bare parameters in the Lagrangian, namely $m_0$ and $g^2$, here we have three independent bare parameters: $\tilde{m}$, $\tilde{g}_\sigma$ and $\tilde{m}_\sigma$. Hence, additional renormalization conditions are required.

It is convenient to turn to counter term renormalization. Introducing the respective counter terms and again integrating out the meson fields, one obtains the renormalized version of Eq. (3.4)

$$\mathcal{L} = \bar{\psi}\left(i\slashed{\partial} - m - \delta_m\right)\psi + \frac{g_\sigma^2}{2}\bar{\psi}\psi \frac{1}{(1+\alpha_\sigma)\Box + m_\sigma^2 - \alpha_2}\bar{\psi}\psi$$
$$+ \frac{g_\omega^2}{2}\bar{\psi}\gamma^\nu\psi \frac{1}{(1+\alpha_\omega)\Box + m_\omega^2}\left(-\delta_\nu^{\,\mu} - \frac{(1+\alpha_\omega)\partial_\nu\partial^\mu}{m_\omega^2}\right)\bar{\psi}\gamma_\mu\psi. \qquad (3.6)$$

From Eq. (3.6), one can easily read off the (momentum space) Feynman propagators, incorporating the appropriate counter terms, for the $\sigma$ and $\omega$ fields. They are given by

$$i\Delta(p) = \frac{i}{(1+\alpha_\sigma)p^2 - m_\sigma^2 + \alpha_2 + i\eta} \qquad (3.7)$$

## 3.1. WALECKA MODEL IN 1+1 DIMENSIONS

and
$$iD_\nu^\mu(p) = \frac{i}{(1+\alpha_\omega)p^2 - m_\omega^2 + i\eta}\left(-\delta_\nu^{\ \mu} + \frac{(1+\alpha_\omega)p_\nu p^\mu}{m_\omega^2}\right), \quad (3.8)$$
respectively.

We first determine the fermion self-energy in the vacuum. As in the previous chapter, the vacuum is assumed to be translationally invariant. In general the full fermion self-energy $\Sigma_H$ (cf. Fig. 3.2) has scalar ($S$) as well as vector ($V_\mu$) contributions

$$\Sigma_H = (S - m) - \gamma^\mu V_\mu. \quad (3.9)$$

However the vector contribution can be shown to vanish in the vacuum, resulting in a purely scalar contribution $\Sigma_H|_\text{vac} = M - m$. Here we introduced the physical fermion mass in the vacuum $M = S|_\text{vac}$. Obviously, the diagrams contributing to the physical fermion mass in the vacuum contain the divergent primitive Feynman diagrams Fig 3.1a and Fig 3.1b. All the other fermion loop diagrams (more than two scalar insertions) are finite in 1+1 dimensions. A common choice [36] is

$$\Sigma_H|_\text{vac} = 0 \quad \leftrightarrow \quad M \equiv m, \quad (3.10)$$

i.e., the mass $m$ occurring in the Lagrangian $\mathcal{L} = \mathcal{L}' + \mathcal{L}_{CT}$ already corresponds to the physical mass in the vacuum. In this sense, all possible loop corrections are already included and $\Sigma_H$ vanishes in the vacuum. It is obvious from Fig. 3.2, that the renormalization condition Eq. (3.10) is fulfilled by fixing $\delta_m$ such that the single primitive Feynman diagram Fig. 3.1a vanishes at zero external momentum transfer

$$\delta_m = -m \frac{N}{\pi} \frac{g_\sigma^2}{m_\sigma^2} \ln \frac{\Lambda}{m}. \quad (3.11)$$

Let us now address the question of how to fix $\alpha_2$. As we have seen above, with the physical fermion mass alone, we do not have enough information to determine both $\delta_m$ and $\alpha_2$. Therefore, an additional observable is necessary. For this reason we turn to the scalar field propagator.

In analogy to the effective coupling $g_\text{eff}^2$ in the GN model (cf. Fig. 2.6, Eq. (2.13)), an effective scalar field propagator $i\Delta'(k)$ can be constructed in the Walecka model. The corresponding diagrams follow from those employed in the GN model, substituting the point interaction of the GN model for an interaction mediated by the scalar meson field. Thus, the two-loop order self-energy is given by

$$\delta\Sigma(k) = -(ig_\sigma)^2 \Delta(k) \langle\bar\psi\psi\rangle_k$$
$$\times \left\{1 + (ig_\sigma)^2 \Delta(k) \frac{N}{\pi}\left(\ln\frac{\Lambda}{m} - 1 + \frac{k^2}{12m^2} + \frac{(k^2)^2}{120m^4} + O(k^6)\right)\right\}. \quad (3.12)$$

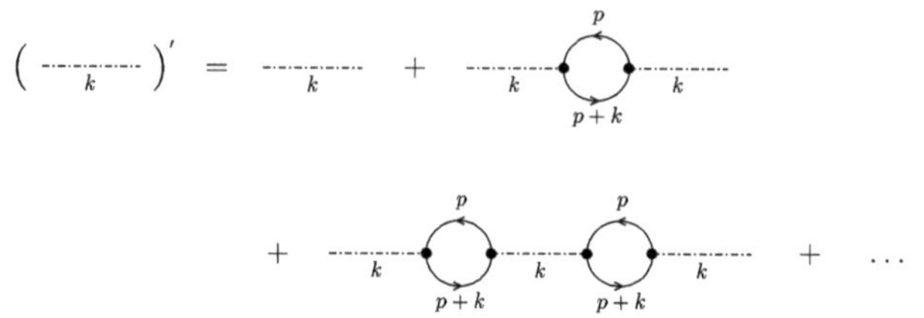

Figure 3.3: Resummation of scalar vacuum polarization graphs coupled via $\sigma$ propagators into an effective scalar field propagator.

We just have to replace the GN coupling constant $g^2$ by the momentum dependent quantity $(ig_\sigma)^2 \Delta(k)$, to obtain the effective scalar interaction for the Walecka model in 1+1 dimensions. Hence, the expression equivalent to Eq. (2.13), after some regrouping of terms, is given by

$$g^2_{\sigma,\text{eff}}(k) = \frac{(ig_\sigma)^2}{\left(1+\alpha_\sigma+\frac{Ng^2_\sigma}{12\pi m^2}\right)k^2 - m^2_\sigma + \alpha_2 + \frac{Ng^2_\sigma}{\pi}\left(\ln\frac{\Lambda}{m}-1\right) + \frac{Ng^2_\sigma}{\pi}\left(\frac{(k^2)^2}{120m^4} + \mathcal{O}(k^6)\right) + i\eta}. \quad (3.13)$$

The $k$-independent divergence can be eliminated by an adequate choice of the counter term $\alpha_2$. We interpret the expression in Eq. (3.13) as an effective scalar propagator, defining $g^2_{\sigma,\text{eff}}(k) = (ig_\sigma)^2 \Delta'(k)$, as illustrated graphically in Fig. 3.3.

Note that both $\alpha_2$ as well as $\alpha_\sigma$ can be determined from $\Delta'(k)$. In the following we turn to a particular choice of the counter terms. Interpreting the parameter $m^2_\sigma$ to be the renormalized, physical mass of the sigma meson in the vacuum, we demand that the effective scalar propagator is of the form

$$i\Delta'(k) = \frac{i}{k^2 - m^2_\sigma - M^2(k^2) + i\eta}, \quad (3.14)$$

where $M^2(k^2)|_{k^2=0} = 0$ and also $\frac{d}{dk^2}M^2(k^2)|_{k^2=0} = 0$. Or, equivalently,

$$\alpha_2 = -\frac{Ng^2_\sigma}{\pi}\left(\ln\frac{\Lambda}{m} - 1\right) \quad (3.15)$$

and

$$\alpha_\sigma = -\frac{Ng^2_\sigma}{12\pi m^2}. \quad (3.16)$$

This fixes all the counter terms associated with the scalar meson field in 1+1 dimensions.

What is still missing is the explicit expression for $\alpha_\omega$, related to the vector field propagator. In order to determine the effective vector field propagator, one has to invoke the same steps as for the effective scalar propagator. One just has to replace the scalar couplings by vector couplings and the scalar propagators by vector propagators in Fig. 3.3. Due to the Lorentz indices,

## 3.1. WALECKA MODEL IN 1+1 DIMENSIONS

the determination of the effective vector propagator is somewhat more involved. Performing the resummation in terms of a geometric series, a matrix in Lorentz indices has to be inverted. To simplify the notation, the unrenormalized " — " loop with $\gamma^\mu$ as well as $\gamma^\nu$ insertion at external momentum transfer $k$ is referred to as $i\Pi^{\mu\nu}(k)$ in the following. Extracting the tensor structure in the Lorentz indices

$$i\Pi^{\mu\nu}(k) = i(g^{\mu\nu}k^2 - k^\mu k^\nu)\Pi(k^2),\qquad(3.17)$$

the vector propagator in the effective theory can be cast into the following algebraic form

$$iD'^{\,\mu}_{\nu}(k) = \frac{i}{k^2 - m_\omega^2 + i\eta}\left[\frac{1}{1 - C(k^2)k^2}(-\delta_\nu^{\,\mu} + C(k^2)k_\nu k^\mu) + \frac{k_\nu k^\mu}{m_\omega^2}\right].\qquad(3.18)$$

Here we have introduced the momentum dependent scalar function

$$C(k^2) = \frac{i}{k^2 - m_\omega^2 + i\eta}\left[g_\omega^2 i\Pi(k^2) + i\alpha_\omega\right].\qquad(3.19)$$

Again the divergent contribution is accompanied by a counter term contribution and hence can be eliminated by an adequate choice of this parameter. Note that for $C(k^2) \equiv 0$, the vector field propagator of the original theory, Eq. (3.8), is recovered.

Let us find a prescription how to fix the counter term $\alpha_\omega$. In principle we proceed as in the scalar case above. We want the effective vector propagator to reduce to $iD_\nu^{\,\mu}(k)$ at small $k$ momentum transfer, i.e.,

$$iD'^{\,\mu}_{\nu}(k) = \frac{i}{k^2 - m_\omega^2 + \mathcal{O}(k^4) + i\eta}\left(-\delta_\nu^{\,\mu} + \frac{k_\nu k^\mu}{m_\omega^2} + \mathcal{O}(k^4)\right).\qquad(3.20)$$

This can be achieved by demanding

$$g_\omega^2 i\Pi(k^2)\big|_{k=0} + i\alpha_\omega = 0\qquad(3.21)$$

or

$$\alpha_\omega = -\frac{Ng_\omega^2}{6\pi m^2}.\qquad(3.22)$$

Finally, all the counter terms required in 1+1 dimension have been determined. While the counter terms multiplying powers of the meson fields ($\delta_m$, $\alpha_2$) are infinite, those multiplying derivatives of the meson fields ($\alpha_\sigma$, $\alpha_\omega$) are finite. The role of the latter is to ensure that the meson propagators have the expected momentum dependence.

We now turn to the explicit construction of the no-sea effective theory for the Walecka model. As in the GN model family, the kinetic part of the no-sea effective Lagrangian is

$$\mathcal{L}_{\text{eff,kin}} = \bar\psi(i\slashed\partial - m)\psi.\qquad(3.23)$$

All the fermion spinors appearing in the remainder of this chapter refer to the " + " sector, hence we will omit the " + " index from now on.

So far, we have identified the effective counter parts of the interactions already present in the full, underlying theory in 1+1 dimensions, Eq. (3.1) ($\tilde{c}_3 = \tilde{c}_4 = 0$). While the interactions of the full fermion field can be thought of as being mediated via the scalar and vector meson field propagators Eqs. (3.7) and (3.8), the corresponding interactions (quadratic in the fermion condensates) between " + " fields in the no-sea effective theory are mediated via the effective propagators Eqs. (3.14) and (3.18),

$$\mathcal{L}^{(2)}_{\text{eff,int}} = -\frac{g_\sigma^2}{2}(\bar{\psi}\psi)\Delta'(\bar{\psi}\psi) - \frac{g_\omega^2}{2}(\bar{\psi}\gamma^\nu\psi)D'_\nu{}^\mu(\bar{\psi}\gamma_\mu\psi)\,, \tag{3.24}$$

in an analogous way. The presence of the propagators in Eq. (3.24) shows that we are dealing with non-local effective interactions between positive energy fermions. Here, as in the previous chapter, our aim is to derive a local no-sea effective theory, valid in the vicinity of the fully occupied Dirac sea only. We restrict ourselves to small external two-momentum transfer $k$ in the effective propagators Eqs. (3.14) and (3.18). This will enable us to derive an "almost local" effective Lagrangian well suited for analytical studies.

Let us therefore trade the effective propagators for effective, $k$-dependent couplings. As both the effective scalar field propagator and the effective vector field propagator describe massive fields, the resulting couplings are well defined in the infrared and do not contain any singularities. We assume that the mass terms dominate the inverse effective meson propagators, i.e., $m_\sigma^2 \gg k^2$ and $m_\omega^2 \gg k^2$, respectively. While the effective scalar coupling is then given by

$$g^2_{\sigma,\text{eff}}(k) = -\mathrm{i}(\mathrm{i}g_\sigma)^2 \mathrm{i}\Delta'(k) = \frac{g_\sigma^2}{m_\sigma^2}\left(1 + \frac{k^2}{m_\sigma^2} + \left[\frac{Ng_\sigma^2}{120m^4\pi} + \frac{1}{m_\sigma^2}\right]\frac{k^4}{m_\sigma^2} + \mathcal{O}(k^6)\right)\,, \tag{3.25}$$

the effective vector coupling is determined by

$$\left(g^2_{\omega,\text{eff}}(k)\right)_\nu{}^\mu = -\mathrm{i}(\mathrm{i}g_\omega)^2 \mathrm{i}D'_\nu{}^\mu(k) = \frac{g_\omega^2}{m_\omega^2}\left(-\delta_\nu{}^\mu - \frac{k^2\delta_\nu{}^\mu - k_\nu k^\mu}{m_\omega^2} + \mathcal{O}(k^4)\right)\,. \tag{3.26}$$

As a result, the non-local, effective interactions in Eq. (3.24) decompose into an infinite number of local interactions (involving increasing powers of derivatives in position space, as $k^\mu \to -\mathrm{i}\partial^\mu$).

Using the language of "effective couplings", we can proceed exactly as we did for the GN model family. Instead of scalar- and pseudoscalar-couplings, we have scalar- and vector-couplings here. However, let us also emphasize the crucial difference between the GN family and the Walecka model. The original four-fermion couplings of the GN model family are point-like, i.e. independent of momentum transfer. Hence, the whole momentum dependence in the effective couplings $g^2_{\text{eff}}(k)$ and $G^2_{\text{eff}}(k)$ (cf. Eqs. (2.14) and (2.52)) is due to the allowance of a small external momentum transfer via the resummed " $-$ " loops.

## 3.1. WALECKA MODEL IN 1+1 DIMENSIONS

In contrast, the original four-fermion interactions of the Walecka model are of finite range and therewith already feature an explicit momentum dependence (cf. Eqs. (3.7) and (3.8)). The origin of the momentum dependence in Eqs. (3.25) and (3.26) is twofold. There are both contributions from the expansion of the original four fermion interaction as well as the resummed " $-$ " loop diagrams. Due to our renormalization prescription, this becomes obvious for $\mathcal{O}(k^4)$ terms only. We focus on the term proportional to $k^4$ of Eq. (3.25). While the first contribution in the square brackets, $\sim g_\sigma^2$, is due to the " $-$ " loop contribution, the latter, $\sim 1/m_\sigma^2$, arises from the $k$-dependent part of the original propagator.

To determine the two-loop effective Lagrangian, we follow closely what we did for the GN model family in the previous chapter. From the scalar self-energy contribution

$$\delta\Sigma(k) = i(ig_{\sigma,\text{eff}}^2)\langle\bar{\psi}\psi\rangle_k$$
$$= -\frac{g_\sigma^2}{m_\sigma^2}\left(1 + \frac{k^2}{m_\sigma^2} + \frac{1}{m_\sigma^2}\left[\frac{Ng_\sigma^2}{120m^4\pi} + \frac{1}{m_\sigma^2}\right]k^4 + \mathcal{O}(k^6)\right)\langle\bar{\psi}\psi\rangle_k, \quad (3.27)$$

we obtain

$$\mathcal{L}_{\text{eff}}^{(2,s)} = \frac{1}{2}\frac{g_\sigma^2}{m_\sigma^2}(\bar{\psi}\psi)^2 - \frac{1}{2}\frac{g_\sigma^2}{m_\sigma^4}(\Box\bar{\psi}\psi)(\bar{\psi}\psi)$$
$$+ \frac{1}{2}\frac{g_\sigma^2}{m_\sigma^4}\left[\frac{1}{m_\sigma^2} + \frac{Ng_\sigma^2}{120m^4\pi}\right](\Box^2\bar{\psi}\psi)(\bar{\psi}\psi) + \cdots. \quad (3.28)$$

The same procedure applied to the vector self-energy contribution yields

$$\mathcal{L}_{\text{eff}}^{(2,v)} = -\frac{1}{2}\frac{g_\omega^2}{m_\omega^2}(\bar{\psi}\gamma_\nu\psi)(\bar{\psi}\gamma^\nu\psi) - \frac{1}{2}\frac{g_\omega^2}{m_\omega^4}\partial^\nu(\bar{\psi}\gamma_\nu\psi)\partial^\mu(\bar{\psi}\gamma_\mu\psi)$$
$$+ \frac{1}{2}\frac{g_\omega^2}{m_\omega^4}(\bar{\psi}\gamma_\nu\psi)(\Box\bar{\psi}\gamma^\nu\psi) + \cdots. \quad (3.29)$$

However, as we know from the GN model, there also arise new interactions which do not have any direct counter parts in the Lagrangian of the full, underlying theory. In the process of determining these, Dirac sea induced, effective interactions, we have to evaluate " $-$ " loop integrals with different numbers of $\gamma^0$, $\gamma^1$ and $1$ insertions. The corresponding interaction terms in the effective Lagrangian can again be inferred from the associated self-energy contributions. The purely scalar contribution to the three-loop Lagrangian is given by

$$\mathcal{L}_{\text{eff}}^{(3,1)} = \frac{N}{6\pi m}\left(\frac{g_\sigma^2}{m_\sigma^2}\right)^3(\bar{\psi}\psi)^3 - \frac{N}{24\pi m^3}\left(\frac{g_\sigma^2}{m_\sigma^2}\right)^3(\Box\bar{\psi}\psi)(\bar{\psi}\psi)^2 + \cdots. \quad (3.30)$$

We already performed the corresponding calculation in the GN model. One simply has to replace

$$(g^2)^3 \to g_{\sigma,\text{eff}}^2(k_1)g_{\sigma,\text{eff}}^2(k_2)g_{\sigma,\text{eff}}^2(k_1 + k_2) \quad (3.31)$$

in Eq. (2.17).

Let us complete the construction of the three-loop effective Lagrangian. We first turn to the effective interaction, arising from " $-$ " loops with two 1 and one $\gamma^\mu$ insertion. Here, it is important to realize that it does not suffice to consider just one diagram, as there is a topological distinct second diagram mimicking the same interaction. The full effective interaction is given as the sum of the two diagrams and vanishes (cf. appendix B.3). The same is true for the effective interaction descending from " $-$ " loops with $\gamma^\mu$, $\gamma^\nu$ and $\gamma^\sigma$ insertions, which vanishes as well (cf. appendix B.3).

What remains, is the effective interaction induced by the " $-$ " loop with $\gamma^\mu$, $\gamma^\nu$ and 1 insertion. In this case, the scalar self-energy is given by

$$\delta\Sigma(k) = i \int \frac{dk_1}{2\pi} \frac{dk_2}{2\pi} (2\pi)\delta(k-k_1-k_2) \left(ig^2_{\sigma,\text{eff}}(k_1+k_2)\right) \left(ig^2_{\omega,\text{eff}}(k_1)\right)^\alpha_\nu \left(ig^2_{\omega,\text{eff}}(k_2)\right)^\beta_\mu$$
$$\times 2\left(-\frac{N}{6\pi m^3}\right) \left(g^{\mu\nu}k_1 k_2 - k_1^\mu k_2^\nu + \mathcal{O}(k^4)\right) \langle\bar\psi\gamma_\alpha\psi\rangle_{k_1} \langle\bar\psi\gamma_\beta\psi\rangle_{k_2}, \qquad (3.32)$$

where the factor 2 accounts for the second contributing diagram ($\nu \leftrightarrow \mu$, $k_1 \leftrightarrow k_2$). Treating both $k_1^\mu$ and $k_2^\mu$ as of $\mathcal{O}(k)$, we obtain

$$\mathcal{L}^{(3,2)}_{\text{eff}} = -\frac{N}{3\pi m^3} \frac{g^2_\sigma}{m^2_\sigma} \left(\frac{g^2_\omega}{m^2_\omega}\right)^2$$
$$\times \bar\psi\psi \left[\partial^\beta\left(\bar\psi\gamma_\alpha\psi\right)\partial_\beta\left(\bar\psi\gamma^\alpha\psi\right) - \partial^\beta\left(\bar\psi\gamma_\alpha\psi\right)\partial^\alpha\left(\bar\psi\gamma_\beta\psi\right)\right] + \cdots. \qquad (3.33)$$

The full three-loop effective Lagrangian is given by $\mathcal{L}^{(3)}_{\text{eff}} = \mathcal{L}^{(3,1)}_{\text{eff}} + \mathcal{L}^{(3,2)}_{\text{eff}}$.

Along these lines, an increasing number of effective interactions can be constructed. In the course of this, it is important to note that no new non-derivative interactions involving higher powers of the vector condensate $\bar\psi\gamma^\nu\psi$ are induced. It can be shown that all " $-$ " loops with at least one $\gamma^\nu$ insertion (and an arbitrary number of 1 insertions) vanish at zero external momentum transfer (cf. appendix B.2). Hence, the only three-loop contribution, which survives when demanding translational invariance (i.e., restricting to zero momentum transfer in the " $-$ " loop diagrams), is the first term in Eq. (3.30).

We do not continue in a fully systematic way beyond this level. However, in view of the applications we have in mind, we single out specific higher order terms.

Let us consider the non-derivative scalar contributions proportional to $(\bar\psi\psi)^4$ and $(\bar\psi\psi)^5$. Obviously, the diagrams giving rise to the effective $(\bar\psi\psi)^4$ interaction are the same as in the GN model (cf. Fig. 2.5c,d) with bare mass term Eq. (2.43). One just has to replace $g^2_{\text{eff}}$ by $g^2_{\sigma,\text{eff}}$

$$\mathcal{L}^{(4+5,s)}_{\text{eff}} = \frac{N}{8\pi m^2}\left(\frac{g^2_\sigma}{m^2_\sigma}\right)^4 \left[\frac{1}{3} + \frac{N}{\pi}\frac{g^2_\sigma}{m^2_\sigma}\right](\bar\psi\psi)^4. \qquad (3.34)$$

## 3.1. WALECKA MODEL IN 1+1 DIMENSIONS

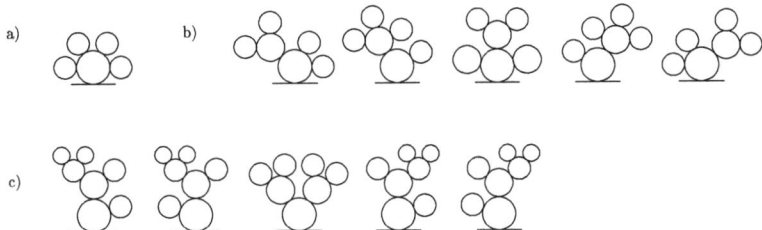

Figure 3.4: Topologically distinct diagrams contributing to the effective $(\bar{\psi}\psi)^5$ interaction. The interaction arises from a) 5-loop, b) 6-loop and c) 7-loop contributions. While the inner loops are " $-$ " loops, the outermost are " $+$ " loops.

In the case of the GN model, this eight-fermion interaction was the highest interaction term considered in the effective Lagrangian. However, since we later on want to compare the effective Lagrangians of the Walecka model in 1+1 and 3+1 dimensions, we proceed to the term proportional to $(\bar{\psi}\psi)^5$ here. It will turn out to be of particular interest in the Walecka model in 3+1 dimensions. The diagrams contributing to this interaction are depicted in Fig. 3.4, with all couplings given by $g^2_{\sigma,\text{eff}}$. Summing up the respective diagrams, we obtain

$$\mathcal{L}^{(5+6+7,s)}_{\text{eff}} = \frac{N}{20\pi m^3}\left(\frac{g^2_\sigma}{m^2_\sigma}\right)^5\left[\frac{1}{3} + \frac{5N}{\pi}\frac{g^2_\sigma}{m^2_\sigma}\left(\frac{1}{3} + \frac{N}{2\pi}\frac{g^2_\sigma}{m^2_\sigma}\right)\right](\bar{\psi}\psi)^5. \qquad (3.35)$$

Collecting the different contributions, we finally arrive at the following no-sea effective Lagrangian

$$\begin{aligned}\mathcal{L}_{\text{eff}} &= \bar{\psi}(i\slashed{\partial} - m)\psi + \frac{1}{2}\frac{g^2_\sigma}{m^2_\sigma}(\bar{\psi}\psi)^2 - \frac{1}{2}\frac{g^2_\sigma}{m^4_\sigma}(\Box\bar{\psi}\psi)(\bar{\psi}\psi) + \frac{1}{2}\frac{g^2_\sigma}{m^2_\sigma}\left[\frac{1}{m^2_\sigma} + \frac{Ng^2_\sigma}{120m^4\pi}\right](\Box^2\bar{\psi}\psi)(\bar{\psi}\psi) \\ &- \frac{1}{2}\frac{g^2_\omega}{m^2_\omega}(\bar{\psi}\gamma_\nu\psi)(\bar{\psi}\gamma^\nu\psi) - \frac{1}{2}\frac{g^2_\omega}{m^4_\omega}\partial^\nu(\bar{\psi}\gamma_\nu\psi)\partial^\mu(\bar{\psi}\gamma_\mu\psi) + \frac{1}{2}\frac{g^2_\omega}{m^4_\omega}(\bar{\psi}\gamma_\nu\psi)(\Box\bar{\psi}\gamma^\nu\psi) \\ &- \frac{N}{3\pi m^3}\left(\frac{g^2_\sigma}{m^2_\sigma}\right)\left(\frac{g^2_\omega}{m^2_\omega}\right)^2 \bar{\psi}\psi\left[\partial^\beta\left(\bar{\psi}\gamma_\alpha\psi\right)\partial_\beta\left(\bar{\psi}\gamma^\alpha\psi\right) - \partial^\beta\left(\bar{\psi}\gamma_\alpha\psi\right)\partial^\alpha\left(\bar{\psi}\gamma_\beta\psi\right)\right] \\ &+ \frac{N}{6\pi m}\left(\frac{g^2_\sigma}{m^2_\sigma}\right)^3(\bar{\psi}\psi)^3 - \frac{N}{24\pi m^3}\left(\frac{g^2_\sigma}{m^2_\sigma}\right)^3(\Box\bar{\psi}\psi)(\bar{\psi}\psi)^2 \\ &+ \frac{N}{8\pi m^2}\left(\frac{g^2_\sigma}{m^2_\sigma}\right)^4\left[\frac{1}{3} + \frac{N}{\pi}\frac{g^2_\sigma}{m^2_\sigma}\right](\bar{\psi}\psi)^4 \\ &+ \frac{N}{20\pi m^3}\left(\frac{g^2_\sigma}{m^2_\sigma}\right)^5\left[\frac{1}{3} + \frac{5N}{\pi}\frac{g^2_\sigma}{m^2_\sigma}\left(\frac{1}{3} + \frac{N}{2\pi}\frac{g^2_\sigma}{m^2_\sigma}\right)\right](\bar{\psi}\psi)^5 + \cdots. \end{aligned} \qquad (3.36)$$

Some parts of this effective action will be tested below.

## 3.1.2 Testing the no-sea effective theory

We propose to test the no-sea effective Lagrangian for a low density system of fermions with Fermi momentum $k_f$, assuming unbroken translational invariance. Only in this (possibly unphysical) case do we have the exact analytical solution to compare with. Let us first consider the exact solution. Demanding translational invariance, we determine the fermion mass at finite Fermi momentum $k_f$

$$\begin{aligned} M - m &= \frac{Ng_\sigma^2}{\pi m_\sigma^2} \left\{ M - m - M \ln\left(\frac{M}{m}\right) - \int_0^{k_f} dp \frac{M}{\sqrt{p^2 + M^2}} \right\} \\ &= \frac{Ng_\sigma^2}{\pi m_\sigma^2} \left\{ M - m - M \ln\left(\frac{k_f + \sqrt{k_f^2 + M^2}}{m}\right) \right\}. \end{aligned} \quad (3.37)$$

In order to derive Eq. (3.37), one has to go back to Fig. 3.2 and re-evaluate the tadpole diagram with the fermion Hartree propagator, featuring a mass $M$ as well as a constant vector condensate $V^\mu$, at finite Fermi momentum [36]

$$\begin{aligned} iG^H(p) &= iG_-^H(p) + iG_+^H(p), \\ iG_-^H(p) &= \frac{i}{\bar{p} - M + i\eta}, \\ iG_+^H(p) &= -\frac{\pi(\bar{p} + M)}{E(\bar{p})} \delta(\bar{p}^0 - E(\bar{p})) \theta(k_f - |\bar{\mathbf{p}}|), \end{aligned} \quad (3.38)$$

where $\bar{p}^\mu = p^\mu + V^\mu$ and $E(\bar{p}) = \sqrt{\bar{\mathbf{p}}^2 + M^2}$. Eq. (3.38) is the generalization of Eq. (2.11), with fermion Hartree mass $M$, to account for a finite, constant vector condensate $V^\mu$ also. Note that Eq. (3.38) is valid in 1+1 as well as 3+1 dimensions. While the first term $iG_-^H(p)$ represents the fermion Hartree propagator in the vacuum, the second term $iG_+^H(p)$ can be associated with the positive energy contribution. In contrast to the situation in the vacuum, the vector part of $\Sigma_H$ also yields a finite contribution. It is determined as follows

$$V^\mu = \frac{Ng_\omega^2}{m_\omega^2} \int \frac{d^2p}{(2\pi)^2} \, \text{Tr}\left\{\gamma^\mu iG^H(p)\right\} = -\frac{Ng_\omega^2}{\pi m_\omega^2} k_f \, \delta_0^\nu. \quad (3.39)$$

Let us now turn to the effective no-sea Lagrangian. If the condensates are assumed to be translationally invariant, only the non-derivative terms enter and we obtain the following effective Lagrangian

$$\begin{aligned} \mathcal{L}'_{\text{eff}} &= \bar{\psi}(i\partial\!\!\!/ - m)\psi + \frac{g_\sigma^2}{2m_\sigma^2}(\bar{\psi}\psi)^2 - \frac{g_\omega^2}{2m_\omega^2}(\bar{\psi}\gamma_\nu\psi)(\bar{\psi}\gamma^\nu\psi) \\ &+ \frac{N}{6\pi m}\left(\frac{g_\sigma^2}{m_\sigma^2}\right)^3 (\bar{\psi}\psi)^3 + \frac{N}{8\pi m^2}\left(\frac{g_\sigma^2}{m_\sigma^2}\right)^4 \left[\frac{1}{3} + \frac{N}{\pi}\frac{g_\sigma^2}{m_\sigma^2}\right](\bar{\psi}\psi)^4 \\ &+ \frac{N}{20\pi m^3}\left(\frac{g_\sigma^2}{m_\sigma^2}\right)^5 \left[\frac{1}{3} + \frac{5N}{\pi}\frac{g_\sigma^2}{m_\sigma^2}\left(\frac{1}{3} + \frac{N}{2\pi}\frac{g_\sigma^2}{m_\sigma^2}\right)\right](\bar{\psi}\psi)^5 + \cdots. \end{aligned} \quad (3.40)$$

## 3.1. WALECKA MODEL IN 1+1 DIMENSIONS

The Euler-Lagrange equation then allows us to identify the self-energy $\Sigma_H$ via

$$\frac{\partial}{\partial \bar{\psi}} \mathcal{L}'_{\text{eff}} = \left(i\slashed{\partial} - \Sigma_H - m\right)\psi = 0. \tag{3.41}$$

Using the decomposition of the full self-energy into scalar and vector contributions, Eq. (3.9), Eq. (3.41) can be recast into

$$\left(i\slashed{\partial} - M + \gamma^\mu V_\mu\right)\psi = 0. \tag{3.42}$$

The effective mass $M$ is then given by

$$M = m - \frac{g_\sigma^2}{m_\sigma^2}\langle\bar{\psi}\psi\rangle - \frac{N}{2\pi m}\left(\frac{g_\sigma^2}{m_\sigma^2}\right)^3 \langle\bar{\psi}\psi\rangle^2 - \frac{N}{2\pi m^2}\left(\frac{g_\sigma^2}{m_\sigma^2}\right)^4\left[\frac{1}{3} + \frac{N}{\pi}\frac{g_\sigma^2}{m_\sigma^2}\right]\langle\bar{\psi}\psi\rangle^3$$
$$- \frac{N}{4\pi m^3}\left(\frac{g_\sigma^2}{m_\sigma^2}\right)^5\left[\frac{1}{3} + \frac{5N}{\pi}\frac{g_\sigma^2}{m_\sigma^2}\left(\frac{1}{3} + \frac{N}{2\pi}\frac{g_\sigma^2}{m_\sigma^2}\right)\right]\langle\bar{\psi}\psi\rangle^4 + \cdots . \tag{3.43}$$

By construction, the condensate that appears in Eq. (3.43) only refers to the positive energy sector. Its explicit expression can be taken over from the analogous consideration of the GN model, Eq. (2.27). Hence, a low density expansion of $M$ in $k_f$ yields

$$M = m\left[1 - c_\sigma \frac{k_f}{m} - \frac{1}{2}c_\sigma^3\left(\frac{k_f}{m}\right)^2 + \left(\frac{1}{6}c_\sigma - \frac{1}{6}c_\sigma^4 - \frac{1}{2}c_\sigma^5\right)\left(\frac{k_f}{m}\right)^3 \right.$$
$$\left. + \left(\frac{1}{3}c_\sigma^2 + \frac{1}{6}c_\sigma^3 - \frac{1}{12}c_\sigma^5 - \frac{5}{12}c_\sigma^6 - \frac{5}{8}c_\sigma^7\right)\left(\frac{k_f}{m}\right)^4 + \cdots\right], \tag{3.44}$$

where we have defined $c_\sigma = \frac{Ng_\sigma^2}{\pi m_\sigma^2}$. The Taylor expansion of the exact result, Eq. (3.37), reproduces Eq. (3.44). As $\langle\bar{\psi}\gamma^1\psi\rangle = 0$ in the translationally invariant state, the vector self-energy associated with Eq. (3.40) is given by

$$V^\mu = -\frac{g_\omega^2}{m_\omega^2}\langle\psi^\dagger\psi\rangle\delta_0^\mu, \tag{3.45}$$

with

$$\langle\psi^\dagger\psi\rangle = \frac{N}{\pi}\int_0^{k_f} dp = \frac{N}{\pi}k_f, \tag{3.46}$$

determined from the valence particle contribution in Eq. (2.11). Eq. (3.45) coincides with the full vector self-energy, Eq. (3.39). This confirms that the coefficients of the non-derivative-terms in $\mathcal{L}_{\text{eff}}$ have been evaluated correctly.

In contrast to the GN model and the massless NJL model considered in the last chapter, no analytical multi particle Hartree solutions are known for the Walecka model. Therefore we cannot test the derivative terms in the present case.

### 3.1.3 Application of no-sea effective theory

So far, we have derived the no-sea effective theory for the 1+1 dimensional Walecka model and tested it in a special case. Here we use our effective Lagrangian to study bound states of $n \leq N$ fermions. Unlike in the GN model where exact bound states are known since the work of DHN, here we have no known results to compare with. Our main focus will be on the question of identifying an expansion parameter which allows us to perform consistent and systematic approximations to the full problem.

We start with the assumption that the leading order interaction term in the effective theory is given by the two non-derivative four-fermion interactions

$$\mathcal{L}_{\text{eff}} = \bar\psi(i\slashed{\partial} - m)\psi + \frac{g_\sigma^2}{2m_\sigma^2}(\bar\psi\psi)^2 - \frac{g_\omega^2}{2m_\omega^2}(\bar\psi\gamma_\nu\psi)(\bar\psi\gamma^\nu\psi). \tag{3.47}$$

In order to find a $n$-fermion bound state solution associated with the no-sea effective Lagrangian (3.47), it is convenient to make use of an approach introduced by Lee et al. [16] to construct exact localized solutions of 1+1 dimensional field theories with four-fermion interactions.

If we assume that the number of bound fermions $n$ does not exceed $N$, all fermions can be filled into a single positive energy level and the HF-equation derived from (3.47) is a non-linear Dirac equation. By a slight generalization of the work of Lee et al., this problem can be solved analytically. This then enables us to organize and take into account higher order corrections to $\mathcal{L}_{\text{eff}}$ in terms of a small parameter, at least in a certain regime.

To apply the method of Lee et al. to the Lagrangian (3.47), we have to generalize their approach to the simultaneous appearance of scalar and vector interactions, featuring two different coupling constants. The derivation can be found in appendix B.4. It turns out that there is a solution only for

$$\frac{g_\sigma^2}{m_\sigma^2} > \frac{g_\omega^2}{m_\omega^2}. \tag{3.48}$$

It remains to specify an expansion parameter. Therefore we invoke additional assumptions. Let us set

$$\frac{1}{m_\sigma^2} = \frac{A}{\gamma m_0^2} \quad \text{and} \quad \frac{1}{m_\omega^2} = \frac{B}{\gamma m_0^2}, \tag{3.49}$$

where $A$, $B$ and $\gamma$ are dimensionless parameters and $m_0$ is a unit mass. (Note that this $\gamma$ has nothing to do with the parameter $\gamma$ defined in the context of the GN model family.) Introducing three independent parameters $A$, $B$ and $\gamma$, the problem is overdetermined. We can set one of the parameters $A$ or $B$ equal to one. This is done such that

$$\max\left(\frac{1}{m_\sigma^2}, \frac{1}{m_\omega^2}\right) = \frac{1}{\gamma m_0^2}, \tag{3.50}$$

## 3.1. WALECKA MODEL IN 1+1 DIMENSIONS

or, equivalently (cf. Eq. (3.49)),
$$\max(A, B) = 1 \tag{3.51}$$
and
$$\min(A, B) = C \leq 1. \tag{3.52}$$
Hence inequality (3.48) translates into
$$C < \frac{g_\sigma^2}{g_\omega^2} \quad \text{or} \quad C > \frac{g_\omega^2}{g_\sigma^2}, \tag{3.53}$$
respectively. Moreover, recall the conditions $m_\sigma^2 \gg k^2$ and $m_\omega^2 \gg k^2$, which were necessary in deriving a local effective theory, featuring effective couplings. Independent of Eq. (3.50), they translate into $\gamma m_0^2 \gg k^2$ and $\gamma m_0^2 C^{-1} \gg k^2$. As $C \leq 1$, both inequalities are simultaneously fulfilled if the single inequality $\gamma m_0^2 \gg k^2$ is true. Here we want to use $\gamma^{-1}$ as our (dimensionless) expansion parameter. This means that we also have to guarantee $\gamma \gg 1$. The specifications on the expansion parameter $\gamma^{-1}$ can thus be summarized by
$$\gamma m_0^2 \gg k^2 \quad \text{and} \quad \gamma \gg 1. \tag{3.54}$$

Incorporating Eq. (3.49) and the appropriate specifications for $A$ and $B$ in the exact $n$-fermion bound state solution of Eq. (3.47), given in appendix B.4, one finds
$$\bar{\psi}\psi \sim \frac{1}{\gamma}, \quad \psi^\dagger \psi \sim \frac{1}{\gamma}, \quad \bar{\psi}\gamma^1\psi = 0, \tag{3.55}$$
the last equation being a direct consequence of our choice of the $\gamma$-matrices. Consequently, we have
$$(\bar{\psi}\psi)^2 \sim \frac{1}{\gamma^2}, \quad (\bar{\psi}\gamma^\nu\psi)^2 \sim \frac{1}{\gamma^2}, \quad \partial_x \sim \frac{1}{\gamma}. \tag{3.56}$$
We are now in a position to count $\gamma^{-1}$-powers of the various terms contributing to the full effective Lagrangian. In accordance with our assumption, interaction terms featuring higher powers of the fermion bilinears are of higher order in $\gamma^{-1}$ and therefore subleading as compared to Eq. (3.47). Let us, for example, concentrate on the Lagrangian valid up to $\mathcal{O}(\gamma^{-7})$
$$\mathcal{L}_{\text{eff}} = \bar{\psi}(i\slashed{\partial} - m)\psi + \frac{g_\sigma^2}{2m_\sigma^2}(\bar{\psi}\psi)^2 - \frac{g_\omega^2}{2m_\omega^2}(\bar{\psi}\gamma_\nu\psi)(\bar{\psi}\gamma^\nu\psi) + \frac{N}{6\pi m}\left(\frac{g_\sigma^2}{m_\sigma^2}\right)^3 (\bar{\psi}\psi)^3$$
$$- \frac{1}{2}\frac{g_\sigma^2}{m_\sigma^4}(\Box\bar{\psi}\psi)(\bar{\psi}\psi) + \frac{1}{2}\frac{g_\omega^2}{m_\omega^4}(\bar{\psi}\gamma_\nu\psi)(\Box\bar{\psi}\gamma^\nu\psi). \tag{3.57}$$

The fermion bound state mass $M_B$ computed from this Lagrangian should be valid up to $\mathcal{O}(\gamma^{-6})$. As $\int dx \sim \gamma$, the validity of the series expansion of $M_B$ is reduced by one power of $\gamma^{-1}$ as compared to the series expansion of $\mathcal{L}_{\text{eff}}$ (cf. Eqs. (2.35) and (2.36)). In order to determine $M_B$, we follow the procedure developed above in the context of the NJL model and use a series expansion in $\gamma^{-1}$. To do this, we turn to the equation of motion derived from Eq. (3.57), written in the following form

$$\left(-\gamma_5 i\partial_x + \gamma^0 S(x) - V_0(x)\right)\psi_\alpha = E_\alpha \psi_\alpha, \qquad (3.58)$$

with

$$S(x) = m + s - \frac{N}{2\pi m}\frac{g_\sigma^2}{m_\sigma^2}s^2 + \frac{1}{m_\sigma^2}\partial_x^2 s \qquad (3.59)$$

and

$$V_0(x) = \rho_0 + \frac{1}{m_\omega^2}\partial_x^2 \rho_0 \qquad (3.60)$$

expressed self-consistently through the condensates

$$s(x) = -\frac{g_\sigma^2}{m_\sigma^2}\langle\bar{\psi}\psi\rangle \quad \text{and} \quad \rho_0(x) = -\frac{g_\omega^2}{m_\omega^2}\langle\psi^\dagger\psi\rangle. \qquad (3.61)$$

Here we used that $\langle\bar{\psi}\gamma^1\psi\rangle$ vanishes (cf. Eq. (3.55)). We solve Eq. (3.58) for a single energy level $\alpha \equiv 0$, assuming $n \leq N$. If $Ag_\sigma^2 - Bg_\omega^2 > 0$, the bound state mass is then given by

$$\frac{M_B}{nm} = 1 - \left(\frac{Ag_\sigma^2 - Bg_\omega^2}{m_0^2}\right)^2 \left\{\frac{n^2}{24}\left(\frac{1}{\gamma}\right)^2 - \frac{n^4}{1920}\left(\frac{Ag_\sigma^2 - Bg_\omega^2}{m_0^2}\right)\left(\frac{9Ag_\sigma^2 - Bg_\omega^2}{m_0^2}\right)\left(\frac{1}{\gamma}\right)^4 \right.$$
$$\left. + \frac{n^4}{180}\left[\frac{N}{\pi}\left(\frac{Ag_\sigma^2}{m_0^2}\right)^3 - 3m^2\left(\frac{Ag_\sigma^2 - Bg_\omega^2}{m_0^2}\right)\left(\frac{A^2g_\sigma^2 - B^2g_\omega^2}{m_0^4}\right)\right]\left(\frac{1}{\gamma}\right)^5\right\}, \qquad (3.62)$$

where we kept terms up to order $1/\gamma^5$. $A$ and $B$ are specified by Eq. (3.50) and below. Note that the effective Lagrangian (3.47) is valid up to $\mathcal{O}(\gamma^{-5})$. Hence the exact fermion bound state mass (B.42) associated with this Lagrangian is valid up to $\mathcal{O}(\gamma^{-4})$. For the self-consistent potentials, we obtain

$$S = m\left\{1 + \frac{s_{22}}{\cosh^2 z}\left(\frac{1}{\gamma}\right)^2 + \left(\frac{s_{42}}{\cosh^2 \xi} + \frac{s_{44}}{\cosh^4 \xi}\right)\left(\frac{1}{\gamma}\right)^4 + \left(\frac{s_{52}}{\cosh^2 \xi} + \frac{s_{54}}{\cosh^4 \xi}\right)\left(\frac{1}{\gamma}\right)^5 \right.$$
$$\left. + \left(\frac{s_{62}}{\cosh^2 \xi} + \frac{s_{64}}{\cosh^4 \xi} + \frac{s_{66}}{\cosh^6 \xi}\right)\left(\frac{1}{\gamma}\right)^6 + \cdots\right\} \qquad (3.63)$$

and

$$V_0 = m\left\{\frac{\rho_{22}}{\cosh^2 z}\left(\frac{1}{\gamma}\right)^2 + \left(\frac{\rho_{42}}{\cosh^2 \xi} + \frac{\rho_{44}}{\cosh^4 \xi}\right)\left(\frac{1}{\gamma}\right)^4 + \left(\frac{\rho_{52}}{\cosh^2 \xi} + \frac{\rho_{54}}{\cosh^4 \xi}\right)\left(\frac{1}{\gamma}\right)^5 \right.$$
$$\left. + \left(\frac{\rho_{62}}{\cosh^2 \xi} + \frac{\rho_{64}}{\cosh^4 \xi} + \frac{\rho_{66}}{\cosh^6 \xi}\right)\left(\frac{1}{\gamma}\right)^6 + \cdots\right\}. \qquad (3.64)$$

The explicit expressions for the coefficients $s_{mn}$, $p_{mn}$ can be found in the appendix B.5.

In this particular example, we have demonstrated that a consistent truncation schema based on an expansion parameter can be found. As a result, we have obtained an analytically computable expression for $M_B$ in powers of $\gamma^{-1}$. As pointed out above, exact bound state solutions are not known for this model so that we have derived new results. They could in principle be checked numerically.

Let us briefly summarize this section. Starting from a fully relativistic quantum field theory, the Walecka model in 1+1 dimensions, we turned to the RHA and derived a renormalized no-sea effective theory for positive energy states only, valid in the vicinity of the completely filled Dirac sea. The main difference to the GN model family is the fact that finite range interactions give rise to additional terms in $\mathcal{L}_{\text{eff}}$ and, more technically, the use of counter term renormalization.

Having achieved this, we focused on a particular parameter regime and therein specified an adequate expansion parameter, which allowed us to organize the different contributions in the effective Lagrangian. Consequently, explicit calculations became feasible. As an application of our effective theory, we analytically determined the bound state mass in a systematic expansion in this parameter.

The next step will be to extend the analysis to 3+1 dimensions and derive a no-sea effective theory of the Walecka model in 3+1 dimensions. This is clearly important if we wish to make contact with more realistic field theories.

## 3.2 Walecka model in 3+1 dimensions

Turning to the Walecka model in 3+1 dimensions, we strongly benefit from the experience gained within its 1+1 dimensional counter part. Apart from minor differences, the procedure to construct the no-sea effective theory is the same as for the Walecka model in 1+1 dimensions. Hence, in the subsequent derivation, we often omit the details and refer to the previous section. Moreover, we have not been able to obtain analytical solutions of $\mathcal{L}_{\text{eff}}$ in 3+1 dimensions. Therefore this chapter is much more compact than the previous one.

Let us start this section by recalling the differences between the 1+1 and 3+1 dimensional versions of Walecka model. As already noted in the beginning of this chapter, there is a larger number of divergent fermion loop diagrams in 3+1 dimensions. All the primitive Feynman diagrams depicted in Fig. 3.1 diverge in 3+1 dimensions. The additional divergent diagrams as compared to 1+1 dimensions are Fig. 3.1c and Fig. 3.1d. Hence, additional counter terms in the Lagrangian are needed.

Note again that in order to generate the respective counter terms, the scalar interactions $\sim \sigma^3$ and $\sim \sigma^4$ in the Lagrangian (3.1) are mandatory. Without incorporating them in the original, unrenormalized Lagrangian, the associated counter terms proportional to $\alpha_3$ and $\alpha_4$

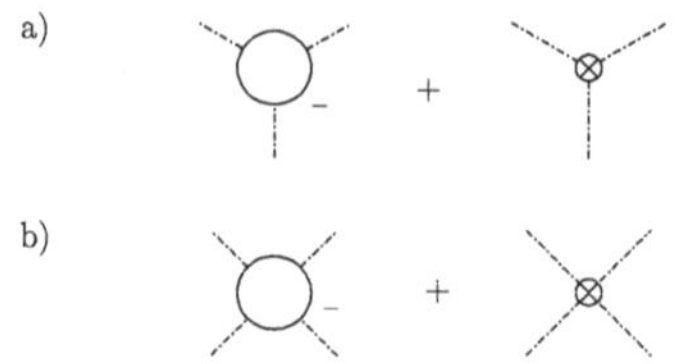

Figure 3.5: Diagrams giving rise to the physical interactions a) $\sim \sigma^3$ and b) $\sim \sigma^4$. The divergent " $-$ " loops are renormalized by an adequate choice of the counter terms $\alpha_3$ and $\alpha_4$.

would not be generated and the divergences in Fig. 3.1c and Fig. 3.1d could not be removed. We can however assume (cf. [36]), that the physical couplings, cubic and quartic in the scalar field, vanish in the vacuum at zero external momentum transfer, i.e., $c_3 = c_4 = 0$. This choice actually corresponds to two renormalization conditions determining the explicit values of the additional counter terms $\alpha_3$ and $\alpha_4$. The physical $\sigma^3$ and $\sigma^4$ interactions are depicted in Fig. 3.5. We fix $\alpha_3$ and $\alpha_4$ in such a way that they exactly cancel the " $-$ " loops with three and four scalar insertions, respectively. Note that non-vanishing physical coefficients $c_3$ and $c_4$ would significantly complicate calculations, but are not required for a consistent model.

There is a further difference concerning the more technical aspect of how to deal with the various (divergent) fermion loop integrals. We turn to dimensional regularization in the following. Unlike in the treatment with a cutoff regularization, all results obtained within this approach are manifestly covariant. Applying this approach, we rely on the following identity

$$\int d^d p \frac{1}{(p^2 + 2pq - m^2)^n} = (-1)^n i\pi^{d/2} \frac{\Gamma\left(n - \frac{d}{2}\right)}{\Gamma(n)} \left(\frac{1}{m^2 + q^2}\right)^{n-d/2}. \quad (3.65)$$

A given integral, divergent in $d = 4$ space-time dimensions, becomes finite when turning to smaller $d$. Regarded as a function of $d$, this finite integral can then be analytically continued into the neighborhood of the physical dimension $d = 4 - \epsilon$, with $\epsilon \to 0$, and the divergence can be isolated. All the other integrals required in this chapter are easily obtained by differentiating Eq. (3.65) with respect to $q_\mu$.

Apart from these differences, the explicit expressions for the effective propagators as well as the respective renormalization conditions for $\delta_m$, $\alpha_2$, $\alpha_\sigma$ and $\alpha_\omega$ can be taken over from $1+1$ dimensions. Due to the presence of the cubic and quartic powers of the scalar meson field, the scalar meson field cannot be integrated out exactly as in the 1+1 dimensional model. However, as in ordinary perturbation theory, one can work with the free renormalized propagators, corresponding to the effective propagators determined in 1+1 dimensions, and the vertices of the theory.

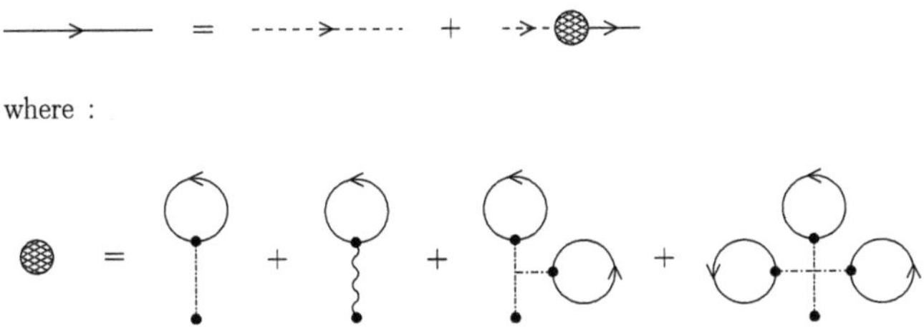

## 3.2. WALECKA MODEL IN 3+1 DIMENSIONS

where :

Figure 3.6: Dyson equation for fermion propagator in RHA. Dashed line: free propagator; solid line: dressed propagator; shaded circle: self-energy; dash-dotted line: scalar propagator; curly line: vector propagator.

For the sake of completeness, the explicit algebraic expressions of the different counter terms, necessary in 3+1 dimensions, are listed below

$$\delta_m = -m^3 \frac{N}{(2\pi)^2} \frac{g_\sigma^2}{m_\sigma^2} \left[ \Gamma\left(-1+\frac{\epsilon}{2}\right) + \ln(m^2\pi) + \mathcal{O}(\epsilon) \right]$$

$$\alpha_2 = -3m^2 \frac{Ng_\sigma^2}{(2\pi)^2} \left[ \Gamma\left(-1+\frac{\epsilon}{2}\right) + \ln(m^2\pi) + \frac{2}{3} + \mathcal{O}(\epsilon) \right]$$

$$\alpha_3 = 6m \frac{Ng_\sigma^3}{(2\pi)^2} \left[ \Gamma\left(-1+\frac{\epsilon}{2}\right) + \ln(m^2\pi) + \frac{5}{3} + \mathcal{O}(\epsilon) \right]$$

$$\alpha_4 = -6 \frac{Ng_\sigma^4}{(2\pi)^2} \left[ \Gamma\left(-1+\frac{\epsilon}{2}\right) + \ln(m^2\pi) + \frac{11}{3} + \mathcal{O}(\epsilon) \right]$$

$$\alpha_\sigma = \frac{1}{2} \frac{Ng_\sigma^2}{(2\pi)^2} \left[ \Gamma\left(-1+\frac{\epsilon}{2}\right) + \ln(m^2\pi) + \frac{5}{3} + \mathcal{O}(\epsilon) \right]$$

$$\alpha_\omega = \frac{1}{3} \frac{Ng_\omega^2}{(2\pi)^2} \left[ \Gamma\left(-1+\frac{\epsilon}{2}\right) + \ln(m^2\pi) + 1 + \mathcal{O}(\epsilon) \right]. \quad (3.66)$$

We start again from the determination of the physical fermion mass in the vacuum. Its self-consistent generation is illustrated in Fig. 3.6. Exactly as in 1+1 dimensions, the fermion tadpole diagram coupled via a vector field vanishes in the vacuum. As in 1+1 dimensions, we assume $m$ to be the physical fermion mass in the vacuum so that the counter terms should completely cancel the self-energy contribution in the vacuum, depicted in Fig. 3.6. Assuming translational invariance, the fermion mass gap equation at finite Fermi momentum $k_f$ can then easily be inferred

$$M - m = \frac{Ng_\sigma^2}{(2\pi)^2 m_\sigma^2} \left\{ -\frac{1}{\pi} \int_0^{k_f} d^3p \frac{M}{\sqrt{p^2 + M^2}} + 2M^3 \ln\left(\frac{M}{m}\right) \right.$$
$$\left. - \frac{11}{3}M^3 + 6mM^2 - 3m^2M + \frac{2}{3}m^3 \right\}. \quad (3.67)$$

Its explicit derivation can be found in [36]. Note that Eq. (3.67) only involves the "static", momentum independent counter terms $\delta_m$, $\alpha_2 \ldots \alpha_4$. While $m$ denotes the fermion mass in the vacuum, $M$ corresponds to the physical fermion mass at finite Fermi momentum $k_f$.

Being interested in a no-sea effective theory valid in the vicinity of the fully occupied Dirac sea only, let us skip the explicit expressions for the effective meson propagators and immediately turn to effective $k$-dependent couplings, assuming $m_\sigma^2 \gg k^2$ and $m_\omega^2 \gg k^2$. While the effective scalar coupling is then given by

$$g_{\sigma,\text{eff}}^2(k) = \frac{g_\sigma^2}{m_\sigma^2}\left\{1 + \frac{1}{m_\sigma^2}k^2 + \frac{1}{m_\sigma^2}\left(\frac{1}{m_\sigma^2} + \frac{1}{80}\frac{Ng_\sigma^2}{\pi^2 m^2}\right)k^4 \right.$$
$$\left. + \frac{1}{m_\sigma^2}\left[\frac{1}{m_\sigma^4} + \frac{1}{40}\frac{Ng_\sigma^2}{\pi^2 m^2}\left(\frac{1}{m_\sigma^2} + \frac{1}{28}\frac{1}{m^2}\right)\right]k^6 + \mathcal{O}(k^8)\right\}, \qquad (3.68)$$

the effective vector coupling is determined by

$$\left(g_{\omega,\text{eff}}^2(k)\right)_\nu^\mu = \frac{g_\omega^2}{m_\omega^2}\left\{-\delta_\nu^\mu - \frac{(\delta_\nu^\mu k^2 - k_\nu k^\mu)}{m_\omega^2}\left[1 + \left(\frac{1}{m_\omega^2} + \frac{1}{60}\frac{Ng_\omega^2}{\pi^2 m^2}\right)k^2 \right.\right.$$
$$\left.\left. + \left(\frac{1}{m_\omega^4} + \frac{1}{10}\frac{Ng_\omega^2}{\pi^2 m^2}\left(\frac{1}{3}\frac{1}{m_\omega^2} + \frac{1}{56}\frac{1}{m^2}\right)\right)k^4\right]\right\} + \mathcal{O}(k^8). \qquad (3.69)$$

The corresponding contributions to the two-loop effective Lagrangian read

$$\mathcal{L}_{\text{eff}}^{(2,s)} = \frac{g_\sigma^2}{2m_\sigma^2}(\bar{\psi}\psi)^2 - \frac{g_\sigma^2}{2m_\sigma^4}(\Box\bar{\psi}\psi)(\bar{\psi}\psi)$$
$$+ \frac{g_\sigma^2}{2m_\sigma^4}\left(\frac{1}{m_\sigma^2} + \frac{1}{80}\frac{Ng_\sigma^2}{\pi^2 m^2}\right)(\Box^2\bar{\psi}\psi)(\bar{\psi}\psi) + \cdots \qquad (3.70)$$

and

$$\mathcal{L}_{\text{eff}}^{(2,v)} = -\frac{g_\omega^2}{2m_\omega^2}(\bar{\psi}\gamma_\nu\psi)(\bar{\psi}\gamma^\nu\psi) + \frac{g_\omega^2}{2m_\omega^4}\left[(\partial_\nu\bar{\psi}\gamma^\nu\psi)(\partial_\mu\bar{\psi}\gamma^\mu\psi) + (\Box\bar{\psi}\gamma_\nu\psi)(\bar{\psi}\gamma^\nu\psi)\right]$$
$$- \frac{g_\omega^2}{2m_\omega^4}\left(\frac{1}{m_\omega^2} + \frac{1}{60}\frac{Ng_\omega^2}{\pi^2 m^2}\right)\left[(\partial_\nu\bar{\psi}\gamma^\nu\psi)(\Box\partial_\mu\bar{\psi}\gamma^\mu\psi) + (\Box^2\bar{\psi}\gamma_\nu\psi)(\bar{\psi}\gamma^\nu\psi)\right] + \cdots, \qquad (3.71)$$

where the ellipses indicate higher derivative terms. In order to determine effective interactions involving higher powers in the fermion condensates, one has to evaluate further " $-$ " loop diagrams, featuring a larger number of 1 as well as $\gamma^\mu$ insertions. The three-loop effective Lagrangian is given by the two non-vanishing contributions

$$\mathcal{L}_{\text{eff}}^{(3,1)} = -\frac{1}{4m}\frac{N}{(2\pi)^2}\left(\frac{g_\sigma^2}{m_\sigma^2}\right)^3(\Box\bar{\psi}\psi)(\bar{\psi}\psi)^2 + \cdots \qquad (3.72)$$

and

$$\mathcal{L}_{\text{eff}}^{(3,2)} = -\frac{1}{3m}\frac{N}{(2\pi)^2}\frac{g_\sigma^2}{m_\sigma^2}\left(\frac{g_\omega^2}{m_\omega^2}\right)^2$$
$$\times \bar{\psi}\psi\left[\partial^\beta(\bar{\psi}\gamma_\alpha\psi)\partial_\beta(\bar{\psi}\gamma^\alpha\psi) - \partial^\beta(\bar{\psi}\gamma_\alpha\psi)\partial^\alpha(\bar{\psi}\gamma_\beta\psi)\right] + \cdots. \qquad (3.73)$$

## 3.2. WALECKA MODEL IN 3+1 DIMENSIONS

Unlike its 1+1 dimensional counter part, the effective Lagrangian of the Walecka model in 3+1 dimensions does not include interactions cubic $\sim (\bar{\psi}\psi)^3$ and quartic $\sim (\bar{\psi}\psi)^4$ in the bilinear $\bar{\psi}\psi$. This is a consequence of our renormalization conditions (cf. Fig. 3.5). The determination of the effective $(\bar{\psi}\psi)^5$-interaction is therefore significantly simpler than in the 1+1 dimensional version of the model. The only contributing diagram is the " $-$ " loop with five 1 insertions. In the effective Lagrangian, it gives rise to the following interaction term

$$\mathcal{L}_{\text{eff}}^{(5)} = -\frac{1}{10m}\frac{N}{(2\pi)^2}\left(\frac{g_\sigma^2}{m_\sigma^2}\right)^5 (\bar{\psi}\psi)^5 + \cdots. \tag{3.74}$$

Combining all of these results, we have so far obtained the following effective Lagrangian of the Walecka model in 3+1 dimensions,

$$\begin{aligned}\mathcal{L}_{\text{eff}} = &\,\bar{\psi}(i\slashed{\partial}-m)\psi + \frac{g_\sigma^2}{2m_\sigma^2}\left(\bar{\psi}\psi\right)^2 - \frac{g_\sigma^2}{2m_\sigma^4}\left(\Box\bar{\psi}\psi\right)(\bar{\psi}\psi) \\ &+\frac{g_\sigma^2}{2m_\sigma^4}\left(\frac{1}{m_\sigma^2}+\frac{1}{80}\frac{Ng_\sigma^2}{\pi^2 m^2}\right)(\Box^2\bar{\psi}\psi)(\bar{\psi}\psi) - \frac{g_\omega^2}{2m_\omega^2}(\bar{\psi}\gamma_\nu\psi)(\bar{\psi}\gamma^\nu\psi) \\ &+\frac{g_\omega^2}{2m_\omega^4}\left[(\partial_\nu\bar{\psi}\gamma^\nu\psi)(\partial_\mu\bar{\psi}\gamma^\mu\psi) + (\Box\bar{\psi}\gamma_\nu\psi)(\bar{\psi}\gamma^\nu\psi)\right] \\ &-\frac{g_\omega^2}{2m_\omega^4}\left(\frac{1}{m_\omega^2}+\frac{1}{60}\frac{Ng_\omega^2}{\pi^2 m^2}\right)\left[(\partial_\nu\bar{\psi}\gamma^\nu\psi)(\Box\partial_\mu\bar{\psi}\gamma^\mu\psi) + (\Box^2\bar{\psi}\gamma_\nu\psi)(\bar{\psi}\gamma^\nu\psi)\right] \\ &-\frac{1}{3m}\frac{N}{(2\pi)^2}\frac{g_\sigma^2}{m_\sigma^2}\left(\frac{g_\omega^2}{m_\omega^2}\right)^2 \bar{\psi}\psi\left[\partial^\beta(\bar{\psi}\gamma_\alpha\psi)\partial_\beta(\bar{\psi}\gamma^\alpha\psi) - \partial^\beta(\bar{\psi}\gamma_\alpha\psi)\partial^\alpha(\bar{\psi}\gamma_\beta\psi)\right] \\ &-\frac{1}{4m}\frac{N}{(2\pi)^2}\left(\frac{g_\sigma^2}{m_\sigma^2}\right)^3 (\Box\bar{\psi}\psi)(\bar{\psi}\psi)^2 - \frac{1}{10m}\frac{N}{(2\pi)^2}\left(\frac{g_\sigma^2}{m_\sigma^2}\right)^5 (\bar{\psi}\psi)^5 + \cdots. \end{aligned} \tag{3.75}$$

Computations based on this effective theory have not yet been performed, so that we cannot address the issue of the systematic classification of all terms here.

If the condensates are assumed to be spatially uniform, only the non-derivative terms enter and Eq. (3.75) reduces to the following effective Lagrangian

$$\mathcal{L}'_{\text{eff}} = \bar{\psi}\left(i\slashed{\partial}-m\right)\psi + \frac{g_\sigma^2}{2m_\sigma^2}\left(\bar{\psi}\psi\right)^2 - \frac{g_\omega^2}{2m_\omega^2}(\bar{\psi}\gamma_\nu\psi)(\bar{\psi}\gamma^\nu\psi) \\ -\frac{1}{10m}\frac{N}{(2\pi)^2}\left(\frac{g_\sigma^2}{m_\sigma^2}\right)^5 (\bar{\psi}\psi)^5 + \cdots. \tag{3.76}$$

The corresponding effective mass is given by

$$M = m - \frac{g_\sigma^2}{m_\sigma^2}\langle\bar{\psi}\psi\rangle + \frac{1}{2m}\frac{N}{(2\pi)^2}\left(\frac{g_\sigma^2}{m_\sigma^2}\right)^5 \langle\bar{\psi}\psi\rangle^4 + \cdots. \tag{3.77}$$

Using the familiar decomposition of the full fermion propagator (mass $M$) at finite Fermi momentum $k_f$, into vacuum and valence particle contributions Eq. (3.38), the explicit expression for the desired condensate can be easily inferred

$$\langle \bar{\psi}\psi \rangle = \frac{2N}{(2\pi)^3} \int_0^{k_f} d^3p \, \frac{M}{\sqrt{\mathbf{p}^2 + M^2}} . \tag{3.78}$$

This implies that the terms written explicitly in Eq. (3.77) should allow the determination of the effective mass valid up to $\mathcal{O}(k_f^{14})$. A low density expansion of $M$ (expanding in $k_f$) then yields

$$M = m - \frac{c_\sigma}{3}(k_f)^3 + \frac{c_\sigma}{10m^2}(k_f)^5 - \frac{3c_\sigma}{56m^4}(k_f)^7 + \frac{c_\sigma^2}{15m^3}(k_f)^8 + \frac{5c_\sigma}{144m^6}(k_f)^9$$
$$- \frac{16c_\sigma^2}{175m^5}(k_f)^{10} + \frac{c_\sigma}{m^4}\left(\frac{c_\sigma^2}{30} - \frac{35}{1408m^4}\right)(k_f)^{11} + \cdots , \tag{3.79}$$

where

$$c_\sigma = \frac{Ng_\sigma^2}{\pi^2 m_\sigma^2} . \tag{3.80}$$

Note that the definition (3.80) is slightly different from the analogous quantity denoted by the same name in 1+1 dimensions, Eq. (3.44). An expansion of the full fermion mass $M$ in Eq. (3.67), determined at finite Fermi momentum $k_f$, assuming translational invariance, reproduces Eq. (3.79) to the order shown.

For the vector self-energy we obtain

$$V^\nu = -\frac{g_\omega^2}{m_\omega^2}\langle \bar{\psi}\gamma^\nu\psi \rangle = -\frac{g_\omega^2}{m_\omega^2}\langle \psi^\dagger\psi \rangle \delta_0^\nu , \tag{3.81}$$

with

$$\rho_B = \langle \psi^\dagger\psi \rangle = 2N \int_0^{k_f} \frac{d^3p}{(2\pi)^3} = \frac{N}{3\pi^2}(k_f)^3 \tag{3.82}$$

determined from the valence particle contribution in Eq. (3.38). In the last step of Eq. (3.81), we have used $\langle \bar{\psi}\gamma\psi \rangle = 0$ for the translationally invariant state.

Expression Eq. (3.81) agrees with the full vector self-energy, determined under the assumption of translational invariance. The comparison of Eqs. (3.79) and (3.81) with the series expansion of their exact counterparts serves as a good test of our no-sea effective theory for the special case of unbroken translational invariance.

The analysis of localized bound states in 3+1 dimensions would presumably require numerical methods and is outside the scope of the present work.

We rather take a step back and discuss several more general features of the no-sea effective theory approach in the context of the Walecka model. In particular we confront our approach with the mean-field theory approach.

## 3.3 General observations and results

### 3.3.1 Uniform nuclear matter

Let us first focus on (uniform) nuclear matter. Keeping only the leading order interaction terms in the effective Lagrangian for spatially uniform matter, we obtain

$$\mathcal{L}'_{\text{eff}} = \bar{\psi}\left(i\slashed{\partial} - m\right)\psi + \frac{g_\sigma^2}{2m_\sigma^2}\left(\bar{\psi}\psi\right)^2 - \frac{g_\omega^2}{2m_\omega^2}(\bar{\psi}\gamma_\nu\psi)(\bar{\psi}\gamma^\nu\psi)\,. \tag{3.83}$$

The effective mass and the effective vector self-energy are then given by

$$M = m - \frac{g_\sigma^2}{m_\sigma^2}\langle\bar{\psi}\psi\rangle\,,$$

$$V^\nu = -\frac{g_\omega^2}{m_\omega^2}\langle\psi^\dagger\psi\rangle\delta_0^\nu\,, \tag{3.84}$$

and the associated equations of motion can be written as

$$\left[i\slashed{\partial} - \left(m - \frac{g_\sigma^2}{m_\sigma^2}\langle\bar{\psi}\psi\rangle\right) - \gamma_0\frac{g_\omega^2}{m_\omega^2}\langle\psi^\dagger\psi\rangle\right]\psi = 0\,. \tag{3.85}$$

Note that Eqs. (3.84) and (3.85) are the defining equations of the mean-field theory (MFT) approach to stationary, uniform nuclear matter ([36], chap. 3.2). Let us however emphasize that the MFT approach significantly differs from our no-sea effective theory approach. It is therefore instructive to summarize the main features of the two approaches.

MFT can be seen as an ad-hoc prescription to gain practical solutions for the complicated field theoretic problem, posed by the Lagrangian (3.1). While the fermion field is still considered as a field operator, the meson fields are assumed to be classical fields with finite values. The Dirac sea, occupied with an infinite number of negative energy fermion states, is simply omitted. In result, there do not occur any divergences and the "bare parameters" in Eq. (3.1) are directly interpreted as their physical counter parts. From a field theoretic point of view, such a drastic procedure seems hard to justify. However, the MFT approach turned out to describe many physical phenomena surprisingly well.

In contrast, our no-sea effective theory can be traced back to a relativistic Hartree approximation of the full, underlying relativistic quantum field theory, Eq. (3.1). The Dirac sea is treated explicitly and enters the derivation of the effective theory in the form of " − " loops. This gives rise to inevitable divergences, whose treatment requires an adequate renormalization procedure. In the course of this, renormalization conditions have to be specified. The resulting effective theory is dependent on these renormalization conditions. Their choice dictates the explicit form of the various couplings in the no-sea effective theory. We choose them such that the effective quantities resemble their bare counter parts in the vicinity of the fully occupied Dirac sea. Only then, having performed the lengthy derivation of the no-sea effective theory, switching from bare to physical parameters, the correspondence noted above can be established. Our no-sea effective theory approach indicates why (as long as a Hartree treatment is justified) the MFT approach yields reasonable results for uniform nuclear matter at small Fermi momentum in the Walecka model.

Note that the no-sea effective theory Lagrangian (3.83) should be valid up to $\mathcal{O}(k_f^{14})$ in 3+1 dimensions, the lowest order correction term being

$$\delta\mathcal{L}'_{\text{eff}} = -\frac{1}{10m}\frac{N}{(2\pi)^2}\left(\frac{g_\sigma^2}{m_\sigma^2}\right)^5 (\bar{\psi}\psi)^5 \sim k_f^{15}, \qquad (3.86)$$

and up to $\mathcal{O}(k_f^2)$ only in 1+1 dimensions, where

$$\delta\mathcal{L}'_{\text{eff}} = \frac{N}{6\pi m}\left(\frac{g_\sigma^2}{m_\sigma^2}\right)^3 (\bar{\psi}\psi)^3 \sim k_f^3. \qquad (3.87)$$

The discrepancy in the order of the correction terms in 1+1 and 3+1 dimensions has two reasons. Trivially, the fermion density is $\sim k_f$ in 1+1, while it is $\sim k_f^3$ in 3+1 dimensions. Furthermore, by our renormalization conditions, in 3+1 dimensions we explicitly ruled out physical interactions cubic and quartic in the scalar meson fields. In contrast, we posed no restrictions on these interactions in 1+1 dimensions.

It is easy to derive the energy of uniform matter at finite Fermi momentum within the framework of the no-sea effective theory. One simply has to solve the single particle equations of motion, derived from the respective effective Lagrangian $\mathcal{L}'_{\text{eff}}$, for their energy eigenvalues and integrate them up. Note that in Hartree approximation this corresponds to integrating over the positive energy solutions of the massive Dirac equation (mass $M$), featuring a chemical potential $V_0$. In addition one has to account for double counting corrections. One obtains

$$\mathcal{E} = 2N\int_0^{k_f}\frac{\mathrm{d}^3 p}{(2\pi)^3}\left(E(p) - V_0\right) + \text{double counting corrections}, \qquad (3.88)$$

where $E(p) = \sqrt{\mathbf{p}^2 + M^2}$, with $M$ from Eq. (3.43) in $d=1$ and Eq. (3.77) in $d=3$ spatial dimensions, respectively.

Let us focus on the model in $3+1$ dimensions. Here Eq. (3.88) can be written as

$$\mathcal{E} = 2N\int_0^{k_f}\frac{\mathrm{d}^3 p}{(2\pi)^3}E(p) + \frac{g_\sigma^2}{2m_\sigma^2}\langle\bar{\psi}\psi\rangle^2 + \frac{g_\omega^2}{2m_\omega^2}\langle\psi^\dagger\psi\rangle^2$$
$$-\frac{2}{5m}\frac{N}{(2\pi)^2}\left(\frac{g_\sigma^2}{m_\sigma^2}\right)^5\langle\bar{\psi}\psi\rangle^5 + \cdots, \qquad (3.89)$$

where we utilized Eq. (3.82). Note the sign-flip of the vector contribution relative to the leading scalar contribution in the energy density (cf. Eq. (3.76)). While the first three terms

## 3.3. GENERAL OBSERVATIONS AND RESULTS

in Eq. (3.89) constitute the full MFT result, the last term written explicitly is the first non-vanishing no-sea effective theory correction ($\sim k_f^{15}$). Since the lowest order term omitted in Eq. (3.89) is proportional to $\langle \bar{\psi}\psi \rangle^6 \sim k_f^{18}$, the terms written explicitly should reproduce the exact result up to $\mathcal{O}(k_f^{17})$. The MFT result should be valid up to $\mathcal{O}(k_f^{14})$.

These predictions can indeed be verified by turning to the series expansion of the exact result for the energy of uniform matter, obtained by performing the full RHA calculation in 3+1 dimensions [36]

$$\mathcal{E}_{\text{RHA}} = 2N \int_0^{k_f} \frac{\mathrm{d}^3 p}{(2\pi)^3} \sqrt{\mathbf{p}^2 + M^2} + \frac{m_\sigma^2}{2g_\sigma^2}(M-m)^2 + \frac{g_\omega^2}{2m_\omega^2}\rho_B^2$$
$$- \frac{N}{2(2\pi)^2}\left\{ M^4 \ln\left(\frac{M}{m}\right) - \frac{25}{12}M^4 + 4mM^3 - 3m^2 M^2 + \frac{4}{3}m^3 M - \frac{1}{4}m^4 \right\}. \quad (3.90)$$

We now turn to the energy per nucleon, determined by $\mathcal{E}/\rho_B$. Obviously, the no-sea effective theory result for $\mathcal{E}/\rho_B$ with $\mathcal{E}$ determined by Eq. (3.89) is valid up to $\mathcal{O}(k_f^{14})$ and can systematically be improved, while the MFT result can be trusted up to $\mathcal{O}(k_f^{11})$ only.

The series expansion of the no-sea effective theory result up to $\mathcal{O}(k_f^{14})$ is given by

$$\frac{\mathcal{E}}{\rho_B} = m + \left[\frac{3(k_f)^2}{10m} - \frac{3(k_f)^4}{56m^3} + \frac{(k_f)^6}{48m^5} - \frac{15(k_f)^8}{1408m^7} + \frac{21(k_f)^{10}}{3328m^9} - \frac{21(k_f)^{12}}{5120m^{11}} + \frac{99(k_f)^{14}}{34816m^{13}} + \cdots\right]$$
$$+ \frac{g_\omega^2}{2m_\omega^2}\rho_B - \frac{g_\sigma^2}{2m_\sigma^2}\rho_B$$
$$+ \frac{g_\sigma^2 \rho_B}{m_\sigma^2 m}\left[\frac{3(k_f)^2}{10m} - \frac{36(k_f)^4}{175m^3} + \frac{16(k_f)^6}{105m^5} - \frac{64(k_f)^8}{539m^7} + \frac{96(k_f)^{10}}{1001m^9} + \cdots\right]$$
$$+ \left(\frac{g_\sigma^2 \rho_B}{m_\sigma^2 m}\right)^2 \left[\frac{3(k_f)^2}{10m} - \frac{351(k_f)^4}{700m^3} + \frac{8803(k_f)^6}{14000m^5} - \frac{305923(k_f)^8}{431200m^7} + \cdots\right]$$
$$+ \left(\frac{g_\sigma^2 \rho_B}{m_\sigma^2 m}\right)^3 \left[\frac{3(k_f)^2}{10m} - \frac{69(k_f)^4}{70m^3} + \cdots\right]$$
$$+ \left(\frac{g_\sigma^2 \rho_B}{m_\sigma^2 m}\right)^4 \left[\frac{g_\sigma^2}{m_\sigma^2}\frac{Nm^3}{40\pi^2} + \frac{3(k_f)^2}{10m}\left(1 - \frac{g_\sigma^2}{m_\sigma^2}\frac{Nm^2}{8\pi^2}\right) + \cdots\right]. \quad (3.91)$$

Here, we take over the grouping, as well as the interpretation of the different contributions from [35], Eq. (4.8). The first term is the fermion rest mass, followed by the non-relativistic Fermi-gas energy and the first few relativistic corrections. The next two terms (proportional to $\rho_B$) give the non-relativistic limit of the potential energy coming from the vector and scalar mesons. The following term in brackets (with overall factor $\rho_B$) is a relativistic correction to the scalar potential energy that arises from the Lorentz contraction factor in the scalar density, evaluated for fermions of mass $m$. The next three terms (with overall factors $\rho_B^2$, $\rho_B^3$ and $\rho_B^4$) are also corrections to the scalar potential energy.

The full MFT and RHA results for the energy per nucleon in uniform nuclear matter with $N = 2$ are depicted together with the no-sea effective theory results valid up to $\mathcal{O}(k_f^{11})$ and

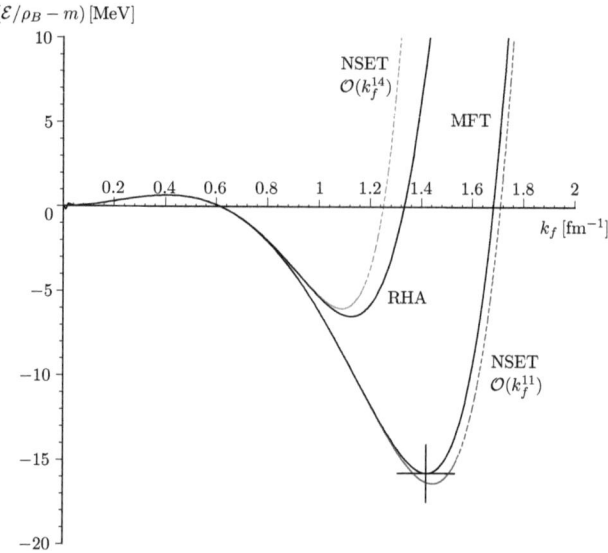

Figure 3.7: The energy per nucleon in uniform nuclear matter with $N=2$ in 3+1 dimensions determined in full MFT and RHA, respectively, is confronted with the no-sea effective theory (NSET) results valid up to $\mathcal{O}(k_f^{11})$ and $\mathcal{O}(k_f^{14})$. The coupling constants are chosen to fit the value and position of the minimum of the MFT curve to the phenomenological values $k_f^0 = 1.42\,\text{fm}^{-1}$, $(\mathcal{E}/\rho_B - m) = -15.75\,\text{MeV}$ [36].

$\mathcal{O}(k_f^{14})$ in Fig. 3.7. For small Fermi momenta $k_f \lesssim 0.9\,\text{fm}^{-1}$ all the curves agree nicely but start to deviate for larger values of $k_f$. The series expansions of the MFT and RHA results agree with each other up to $\mathcal{O}(k_f^{11})$. By including higher order interaction terms in the no-sea effective Lagrangian, the no-sea effective theory result can be systematically improved to reproduce the RHA result up to any desired order. The higher order terms constituting the full MFT result cannot be justified in the no-sea effective theory approach.

While the MFT and RHA curves share the same general characteristics, namely a local minimum and a steeply rising behavior towards larger Fermi momenta, the positions of the minima differ by a Fermi momentum of $\cong 0.25\,\text{fm}^{-1}$ and an energy $\cong 10\,\text{MeV}$ (cf. [36]). Including the first no-sea effective theory correction term $\sim (\bar{\psi}\psi)^5$ in Eq. (3.89), the corresponding curve valid up to $\mathcal{O}(k_f^{14})$ clearly approaches the RHA curve.

## 3.3.2 Spatially non-uniform systems

Let us extend the analysis of uniform nuclear matter in its ground state to include spatially inhomogeneous systems. This is necessary to describe finite "nuclei" or bound states. While the MFT approach to stationary, spatially uniform matter can be considered as leading order treatment within our no-sea effective theory approach, the situation is different for spatially non-uniform systems. Here, the respective MFT equations cannot be inferred directly from our no-sea effective Lagrangian. In order to clarify the differences, let us shortly review the MFT approach to spatially non-uniform systems in the context of the Walecka model in 3+1 dimensions.

The starting point of this approach are the field equations of the original Lagrangian, Eq. (3.1). While the fermion field is still considered as a field operator, the meson fields are replaced by their expectation values, which are assumed to be classical fields with finite values. The Dirac sea, occupied with an infinite number of negative energy fermion states, is simply omitted. In result, there do not occur any divergences and the "bare parameters" in Eq. (3.1) are directly interpreted as their physical counter parts. As above, we do not allow for physical couplings cubic and quartic in the scalar meson field and hence assume $\tilde{c}_3 = \tilde{c}_4 = 0$. In result, the Euler-Lagrange equation for $\psi$ becomes

$$[i\gamma_\mu \partial^\mu - g_\omega \gamma_\mu \langle \omega^\mu \rangle - (m - g_\sigma \langle \sigma \rangle)]\psi = 0 \,. \quad (3.92)$$

The classical meson fields satisfy the field equations

$$(\nabla^2 - m_\sigma^2)\langle \sigma \rangle = -g_\sigma \langle \bar{\psi}\psi \rangle \,, \quad (3.93)$$

$$(\nabla^2 - m_\omega^2)\langle \omega^0 \rangle = -g_\omega \langle \psi^\dagger \psi \rangle \,. \quad (3.94)$$

Since one considers matter at rest, the classical three vector field $\langle \boldsymbol{\omega} \rangle$ is assumed to vanish [35]. Hence, the classical meson fields are generated self-consistently by the fermion condensates.

We formally rewrite Eqs. (3.93) and (3.94) as

$$\langle \sigma \rangle = -g_\sigma \frac{1}{\nabla^2 - m_\sigma^2} \langle \bar{\psi}\psi \rangle \,,$$

$$\langle \omega^0 \rangle = -g_\omega \frac{1}{\nabla^2 - m_\omega^2} \langle \psi^\dagger \psi \rangle \,, \quad (3.95)$$

and insert them in Eq. (3.92). This yields the following equation of motion

$$\left[i\gamma_\mu \partial^\mu + \gamma_0 g_\omega^2 \frac{1}{\nabla^2 - m_\omega^2} \langle \psi^\dagger \psi \rangle - \left(m + g_\sigma^2 \frac{1}{\nabla^2 - m_\sigma^2} \langle \bar{\psi}\psi \rangle \right)\right] \psi = 0 \,. \quad (3.96)$$

It is important to note that while the original field equations obviously refer to the full fermion field $\psi$, in MFT one uses them as equations for $\psi_+$ only, simply substituting $\psi$ for $\psi_+$.

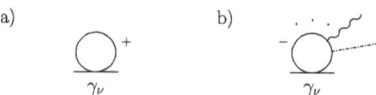

Figure 3.8: Illustration of the effective fermion current $\langle\bar\psi\gamma_\nu\psi\rangle$. a) Valence part $\langle\bar\psi\gamma_\nu\psi\rangle_+$. b) Residual contribution to the fermion current starting with a " $-$ " loop. It is given by the infinite sum of topologically distinct diagrams, made up of further internal " $-$ " loops and outermost " $+$ " loops, coupled via $\sigma$ and $\omega$ propagators.

At first sight, Eq. (3.96) significantly differs from any equations of motion derived from our no-sea effective theory above. We therefore take a step back and reconsider the situation.

Before turning to an almost local effective theory, explicitly restricting to small momentum transfer and defining effective coupling constants, we shortly considered the respective non-local effective theory (cf. Eq. (3.24)). On this level the scalar self-energy can be written as

$$S(k) = m + g_\sigma^2 \Delta'(k)\langle\bar\psi\psi\rangle_k + \cdots. \qquad (3.97)$$

The vector part is determined analogously

$$V_\nu(k) = -g_\omega^2 D'_\nu{}^\mu(k)\langle\bar\psi\gamma_\mu\psi\rangle_k + \cdots. \qquad (3.98)$$

In the Walecka model the fermion current is conserved

$$\partial^\nu\langle\bar\psi\gamma_\nu\psi\rangle = 0 \quad\leftrightarrow\quad k^\nu\langle\bar\psi\gamma_\nu\psi\rangle_k = 0. \qquad (3.99)$$

Note that $\psi$ in Eqs. (3.99) is the full fermion field. In contrast to uniform nuclear matter, it is not a-priori clear that an analogous relation for " $+$ " states holds for spatially non-uniform systems also (cf. Eq. (2.75) in the context of the NJL$_2$ model). The expression for the fermion current translates into an infinite sum of topologically distinct diagrams (cf. Fig. 3.8). Turning to Eq. (3.99), the momentum $k^\nu$ enters the initial " $-$ " loop, Fig. 3.8b, in the form of $k^\nu\gamma_\nu = \slashed{k}$. It can be shown that the contribution of a given " $-$ " loop, with arbitrary 1 and $\gamma^\mu$ insertions, vanishes if one inserts an additional momentum via $\slashed{k}$. To account for all topologically distinct diagrams, one has to sum over all possible insertion points (cf. [38], p. 238-241).

As a result we obtain

$$\partial^\nu\langle\bar\psi\gamma_\nu\psi\rangle_+ = 0. \qquad (3.100)$$

For stationary systems, Eq. (3.100) reduces to $\nabla\langle\bar\psi\gamma\psi\rangle_+ = 0$. This is in particular fulfilled for $\langle\bar\psi\gamma\psi\rangle_+ = 0$, corresponding to "positive energy matter" at rest.

Hence, Eq. (3.98) reduces to

$$V_\nu(k) = -g_\omega^2 D'_\nu{}^0(k)\langle\psi^\dagger\psi\rangle_k + \cdots. \qquad (3.101)$$

## 3.3. GENERAL OBSERVATIONS AND RESULTS

Transforming into position space and inserting the explicit expressions for the effective meson propagators, assuming stationary condensates, we obtain

$$S(x) = m + g_\sigma^2 \frac{1}{\nabla^2 - m_\sigma^2 + \mathcal{O}\left((\nabla^2)^2\right)} \langle \bar\psi \psi \rangle + \cdots \qquad (3.102)$$

and

$$V^0(x) = g_\omega^2 \frac{1}{\nabla^2 - m_\omega^2 + \mathcal{O}\left((\nabla^2)^2\right)} \langle \psi^\dagger \psi \rangle + \cdots, \qquad (3.103)$$

which enter the equation of motion for the "+" fermion field

$$\left(i\slashed\partial + \gamma_0 V^0(x) - S(x)\right)\psi = 0. \qquad (3.104)$$

Concentrating on Eqs. (3.96) and (3.104), MFT dealing with stationary, non-uniform systems and our no-sea effective theory approach can be compared.

The MFT result can be recovered from our no-sea effective theory approach by omitting both the $\mathcal{O}\left((\nabla^2)^2\right)$ terms in the denominators of Eqs. (3.102) and (3.103), as well as the Dirac sea induced terms of higher loop order, involving increasing powers in the fermion condensates and effective propagators, indicated by the ellipses. In the framework of our no-sea effective theory, we do not see how such an "approximation" could be justified.

To illustrate the differences between the two approaches, we turn to the special case where the meson masses dominate the denominators of the meson propagators. This was assumption in the construction of our no-sea effective Lagrangians. Only then do the interactions become almost local and effective coupling constants can be defined. Going back to the no-sea effective Lagrangian, Eq. (3.75), the equation replacing Eq. (3.104) is

$$\left\{ i\gamma_\mu \partial^\mu - \frac{g_\omega^2}{m_\omega^2}\gamma_0 \left(1 + \frac{\nabla^2}{m_\omega^2} + \left(\frac{1}{m_\omega^2} + \frac{N g_\omega^2}{60\pi^2 m^2}\right)\frac{(\nabla^2)^2}{m_\omega^2} + \cdots\right)\langle\psi^\dagger\psi\rangle \right.$$
$$\left. - \left[m - \frac{g_\sigma^2}{m_\sigma^2}\left(1 + \frac{\nabla^2}{m_\sigma^2} + \left(\frac{1}{m_\sigma^2} + \frac{N g_\sigma^2}{80\pi^2 m^2}\right)\frac{(\nabla^2)^2}{m_\sigma^2} + \cdots\right)\langle\bar\psi\psi\rangle + \cdots\right]\right\}\psi = 0. \qquad (3.105)$$

The final ellipses denote contributions which are of higher powers in the fermion condensates, descending from effective interactions not present in the original Lagrangian. Eq. (3.96) can be written as

$$\left\{ i\gamma_\mu\partial^\mu - \frac{g_\omega^2}{m_\omega^2}\gamma_0\left(1 + \frac{\nabla^2}{m_\omega^2} + \frac{(\nabla^2)^2}{m_\omega^4} + \cdots\right)\langle\psi^\dagger\psi\rangle \right.$$
$$\left. - \left[m - \frac{g_\sigma^2}{m_\sigma^2}\left(1 + \frac{\nabla^2}{m_\sigma^2} + \frac{(\nabla^2)^2}{m_\sigma^4} + \cdots\right)\langle\bar\psi\psi\rangle\right]\right\}\psi = 0. \qquad (3.106)$$

Keeping terms up to $\mathcal{O}(\nabla^2)$ in the expansion of the meson propagators only, Eqs. (3.105) and (3.106) coincide. This is a consequence of our renormalization procedure. We chose $\alpha_2$, $\alpha_\sigma$ and $\alpha_\omega$ such that up to this order the effective meson propagators in the no-sea effective theory resemble their counter parts in the underlying relativistic field theory. Turning to higher order terms, differences are unveiled.

The coefficients of the $(\nabla^2)^2$-derivatives of the condensates $-\frac{g_\omega^2}{m_\omega^2}\langle\psi^\dagger\psi\rangle$ and $-\frac{g_\sigma^2}{m_\sigma^2}\langle\bar\psi\psi\rangle$ differ by contributions proportional to the square of the coupling constants. Moreover, the effective multi-fermion interactions, arising in the construction of our no-sea effective theory, are not accounted for within the MFT. Hence, in contrast to the MFT approach to uniform matter, in general the MFT approach to non-uniform systems cannot be seen as a systematic leading order treatment within the framework of the no-sea effective theory.

We believe that the no-sea effective theory approach provides us with a new tool to account for the effects of the Dirac sea in a controllable manner, avoiding ad-hoc approximations. Applications of this method to spatially non-uniform systems will require numerical studies and have not yet been performed.

# Chapter 4
# Conclusions and outlook

Starting from a particular quantum field theoretical model, the GN model family in 1+1 dimensions, we have developed an approach to construct a no-sea effective theory from the underlying relativistic QFT. While the Dirac sea is "integrated out", the positive energy fermion states are treated explicitly. Our approach is tailored to the RHA. In the context of the GN model family, all UV divergences can be removed by invoking the fermion mass gap equation in the vacuum. The resulting no-sea effective Lagrangian features an infinite number of effective interaction terms and requires some truncation. After testing the no-sea effective theory approach by confirming known results in the $GN_2$ model and the $NJL_2$ model without bare mass term, we applied it to the $NJL_2$ model with bare mass term, where no exact analytical solutions are known. Here, we obtained new analytical results for the baryon mass, as well as the associated HF condensates in particular parameter regimes. We also addressed the question how the phenomenon of "induced fermion number" can be recovered in a no-sea effective theory and obtained some insights into the role of the "pion" field. Moreover, we clarified why previous "classical" baryon solutions, obtained by simply neglecting the Dirac sea and accounting for one positive energy level only, yielded useful results for the $GN_2$ model but completely failed in case of the $NJL_2$ model. This part of our work has been published already [39, 31].

In the second chapter, we turned to the Walecka model, first in 1+1 dimensions. The basic approach could be carried over from the GN model family, except that we switched to counter term renormalization. Due to the presence of elementary meson fields, the number of physical parameters is significantly larger in the Walecka model. In contrast to the GN model family, the fermion mass gap equation does not exhaust the renormalization of this model. However, the main difference to the GN model family is the fact that finite range interactions in the Walecka model give rise to additional terms in the respective no-sea effective Lagrangian. For the case of uniform nuclear matter our no-sea effective theory could be tested quantitatively. As far as spatially non-uniform systems are concerned, we showed that a consistent truncation scheme based on an expansion parameter can be found. This was demonstrated in one particular example. We obtained an analytical expression for the $n$-fermion bound state mass in powers of the expansion parameter. As exact bound state solutions are not known for this model, these results are new.

Subsequently, we extended the analysis to 3+1 dimensions and derived a no-sea effective theory of the 3+1 dimensional Walecka model. For uniform matter the no-sea effective theory could once again be verified, like in the 1+1 dimensional version of the model. The analysis of localized bound states in 3+1 dimensions requires numerical methods and is outside the scope of this work.

Finally, we confronted our no-sea effective theory approach with the standard MFT approach to the Walecka model, wherein the Dirac sea is simply omitted. It turned out that MFT applied to uniform nuclear matter at small Fermi momentum can be reinterpreted as leading order treatment of the no-sea effective theory. We have determined the order in a series expansion in $k_f$ to which the MFT results can be trusted. For the 3+1 dimensional Walecka model, the energy per nucleon determined by the MFT approach is only valid up to $\mathcal{O}(k_f^{11})$. Within the no-sea effective theory approach, the order of validity can be improved systematically. Extending the analysis to spatially non-uniform systems, the situation turned out to be more complicated. Here we emphasized the significant differences between the two approaches.

Hence using the GN model family in 1+1 dimensions and the Walecka model in 1+1 as well as 3+1 dimensions as examples, we have demonstrated that "integrating out the Dirac sea" is indeed a viable concept. We see its potential as twofold. On the one hand, the no-sea effective theory approach can be considered as an useful tool to provide new analytical insights in particular parameter regimes. On the other hand, the no-sea effective theory approach itself provides new conceptual insights. We learn how a theory featuring positive energy states only does arise from the full underlying QFT and how the Dirac sea is encoded in the couplings and interaction terms of the resulting no-sea effective theory. This also sheds new lights on several problems of phenomenological interest, notably in 3+1 dimensions.

As the derivation of the no-sea effective theory is based on the evaluation of standard Feynman diagrams, our approach is quite general and not restricted to particular space-time dimensions or QFTs. Once renormalization conditions are specified, the effective coupling constants and interactions of the resulting effective theory are well-defined and finite. There are no UV divergences as in the underlying QFT. Within the no-sea effective theory, the determination of physical observables becomes rather easy. To perform explicit calculations however, a small expansion parameter, allowing for the classification of the infinite number of Dirac sea induced, effective interaction terms generated, has to be identified. Due to the issue of self-consistency, in practice this can only be achieved at the end of the calculation. Combining the no-sea effective theory approach with numerical techniques should give us a handle to attack relativistic matter from a new angle. The main limitation of the present approach is the fact that it relies on the mean field approximation. It would be very interesting to go beyond this particular case.

# Appendix A

# Gross-Neveu model family

## A.1 Results for coefficients defined in Sec. 2.2

Here we collect the $\gamma$-dependent coefficients which enter the results of the no-sea effective theory in Sec. 2.2. Coefficients for the scalar potential $S$, Eq. (2.95),

$$s_{22} = -\frac{1}{4(1+\gamma)^2}, \quad s_{42} = \frac{5\gamma^2 + 3\gamma + 2}{96\gamma(1+\gamma)^5}$$

$$s_{44} = -\frac{3\gamma + 4}{64\gamma(1+\gamma)^4}$$

$$s_{62} = -\frac{91\gamma^4 - 102\gamma^3 + 275\gamma^2 + 276\gamma + 88}{23040\gamma^2(1+\gamma)^8}$$

$$s_{64} = \frac{81\gamma^3 + 13\gamma^2 - 140\gamma - 24}{9216\gamma^2(1+\gamma)^7}$$

$$s_{66} = -\frac{45\gamma^2 - 136\gamma - 60}{9216\gamma^2(1+\gamma)^6}. \tag{A.1}$$

Coefficients for the pseudoscalar potential $P$, Eq. (2.95),

$$p_{33} = \frac{1}{8\gamma(1+\gamma)^2}, \quad p_{53} = -\frac{(3\gamma + 2)(\gamma - 3)}{192\gamma^2(1+\gamma)^5}$$

$$p_{55} = \frac{\gamma - 2}{64\gamma^2(1+\gamma)^4}. \tag{A.2}$$

Coefficients for the valence fermion density $\rho_{\text{val}}$, Eq. (2.101),

$$v_{22} = \frac{1}{4(1+\gamma)}, \quad v_{42} = -\frac{\gamma^2 + \gamma + 1}{48\gamma(1+\gamma)^4}$$

$$v_{44} = \frac{\gamma + 4}{64\gamma(1+\gamma)^3}$$

$$v_{62} = \frac{2\gamma^4 + 10\gamma^3 + 49\gamma^2 + 37\gamma + 11}{2880\gamma^2(1+\gamma)^7}$$

$$v_{64} = -\frac{15\gamma^3 + 151\gamma^2 + 76\gamma + 48}{9216\gamma^2(1+\gamma)^6}$$
$$v_{66} = \frac{9\gamma^2 - 40\gamma + 12}{9216\gamma^2(1+\gamma)^5}. \tag{A.3}$$

Coefficients for the induced fermion density $\rho_{\text{ind}}$, Eq. (2.102),

$$i_{42} = \frac{1}{16\gamma(1+\gamma)^3}, \quad i_{44} = -\frac{3}{32\gamma(1+\gamma)^3}$$
$$i_{62} = -\frac{2\gamma^2 - 5\gamma - 4}{192\gamma^2(1+\gamma)^6}, \quad i_{64} = \frac{4\gamma^2 - 7\gamma - 8}{128\gamma^2(1+\gamma)^6}$$
$$i_{66} = -\frac{5(\gamma-2)}{256\gamma^2(1+\gamma)^5}. \tag{A.4}$$

## A.2   GN$_2$ model with counter term renormalization

The GN model family in 1+1 dimensions is conventionally renormalized by use of the fermion mass gap equation in the vacuum, Eq. (2.42). As only logarithmic divergences $\sim \ln \Lambda$ occur in the theory, all physical quantities can be rendered finite by invoking this equation. However, we think that it is instructive to consider the GN model in counter term renormalization also.

The GN$_2$ model is defined by the following Lagrangian

$$\mathcal{L} = \bar{\psi}\left(i\partial\!\!\!/ - m_0\right)\psi + \frac{g^2}{2}\left(\bar{\psi}\psi\right)^2, \tag{A.5}$$

where $m_0$ and $g$ are bare parameters. Rescaling the fermion field

$$\psi = Z^{1/2}\psi_r, \tag{A.6}$$

and introducing the counter term parameters

$$\delta_2 = Z - 1 \tag{A.7}$$
$$\delta_m = Zm_0 - m \tag{A.8}$$
$$\delta_{g^2} = Z^2 g^2 - G^2, \tag{A.9}$$

the Lagrangian is spit into two terms

$$\mathcal{L} = \mathcal{L}' + \mathcal{L}_{CT}. \tag{A.10}$$

While the first contribution is given by

$$\mathcal{L}' = \bar{\psi}_r\left(i\partial\!\!\!/ - m\right)\psi_r + \frac{G^2}{2}\left(\bar{\psi}_r\psi_r\right)^2, \tag{A.11}$$

the latter corresponds to the counter term Lagrangian

$$\mathcal{L}_{CT} = \bar{\psi}_r \left(\delta_2 i\partial\!\!\!/ - \delta_m\right) \psi_r + \delta_{g^2} \frac{1}{2} \left(\bar{\psi}_r \psi_r\right)^2 . \tag{A.12}$$

As the counter term $\delta_2$ is not needed here, we set $\delta_2 \equiv 0$. In result, the Lagrangian (A.10) can be written as

$$\mathcal{L} = \bar{\psi}_r \left(i\partial\!\!\!/ - m\right) \psi_r + \frac{g^2}{2} \left(\bar{\psi}_r \psi_r\right)^2 - \delta_m \bar{\psi}_r \psi_r , \tag{A.13}$$

where $\delta_m = m_0 - m$. We renormalize in the vacuum and hence turn to the fermion mass gap equation in the vacuum. The physical fermion mass $M$ in the vacuum is determined by

$$M - m = M \frac{Ng^2}{\pi} \ln \frac{\Lambda}{M} + \delta_m . \tag{A.14}$$

As we want to reproduce the mass gap equation in the vacuum written in the form of Eq. (2.42), it suggests itself to work with the bare coupling constant $g$ here.

Let us fix the counter term $\delta_m$ such that the right-hand side of Eq. (A.14) vanishes

$$\delta_m = -M \frac{Ng^2}{\pi} \ln \frac{\Lambda}{M} . \tag{A.15}$$

The physical fermion mass in the vacuum is then given by $M = m$. Inserting Eq. (A.15) in Eq. (A.8), the well known fermion mass gap equation in the vacuum is recovered

$$m - m_0 = m \frac{Ng^2}{\pi} \ln \frac{\Lambda}{m} . \tag{A.16}$$

Defining the dimensionless, renormalization group invariant parameter

$$\gamma := \frac{\pi}{Ng^2} \frac{m_0}{m} , \tag{A.17}$$

Eq. (A.16) can be cast into (cf. Eq. (2.42))

$$1 = \frac{Ng^2}{\pi} \left( \ln \frac{\Lambda}{m} + \gamma \right) . \tag{A.18}$$

So far, we have cured the divergence occurring in the primitive Feynman diagram with a single scalar insertion. The primitive Feynman diagram with two scalar insertions is divergent as well. It can be assigned to the physical coupling constant, whose renormalization is carried out in the next step.

We fix $G^2$ such that it corresponds to the physical coupling constant in the vacuum. To achieve this, we have to sum up the scalar vacuum polarization graphs depicted in Fig 2.6 at vanishing external momentum transfer. The physical coupling constant is then given by the following equation (cf. Eq. (2.13))

$$iG^2 = \frac{i\left(G^2 + \delta_{g^2}\right)}{1 - \frac{N\left(G^2 + \delta_{g^2}\right)}{\pi} \left(\ln \frac{\Lambda}{m} - 1\right)} , \tag{A.19}$$

## A.2. GN$_2$ MODEL WITH COUNTER TERM RENORMALIZATION

where we utilized Eq. (A.9) with $Z = 1$. Solving Eq. (A.19) for $\delta_{g^2}$, we obtain

$$\delta_{g^2} = \left[\frac{1}{G^2} + \frac{N}{\pi}\left(\ln\frac{\Lambda}{m} - 1\right)\right]^{-1} - G^2. \quad (A.20)$$

In principle the renormalization has been completed at this point. Both counter terms $\delta_m$ and $\delta_{g^2}$ have been determined. Note however that inserting Eq. (A.20) in Eq. (A.9), the following equation is yielded

$$\frac{g^2}{G^2} + \frac{Ng^2}{\pi}\left(\ln\frac{\Lambda}{m} - 1\right) = 1. \quad (A.21)$$

Hence, invoking the fermion mass gap equation in the vacuum, Eq. (A.18), the physical coupling constant in the vacuum is given by

$$G^2 = \frac{\pi}{N(1+\gamma)}. \quad (A.22)$$

Note that as soon as the physical mass in the vacuum is fixed, the physical coupling constant in the the vacuum is unambiguously determined here. This reflects the fact, that the fermion mass gap equation suffices in the renormalization of this particular model (cf. chapter 2).

# Appendix B

# Walecka model

## B.1 Walecka model with counter term renormalization

Note that in this appendix, we use a slightly different notation as compared to the main text. While we explicitly distinguish between bare and renormalized fields here, to keep notations simple we do not do this in chapter 3. Obviously, as soon as we incorporate the counter term Lagrangian in Eq. (3.3), we work with renormalized fields also.

The Lagrangian of the Walecka model is given by

$$\mathcal{L} = \bar{\psi}\big(\gamma_\mu(i\partial^\mu - \tilde{g}_\omega \omega^\mu) - (\tilde{m} - \tilde{g}_\sigma \sigma)\big)\psi + \frac{1}{2}\big(\partial_\mu \sigma \partial^\mu \sigma - \tilde{m}_\sigma^2 \sigma^2\big)$$
$$+ \frac{1}{3!}\tilde{c}_3 \sigma^3 + \frac{1}{4!}\tilde{c}_4 \sigma^4 - \frac{1}{4}F_{\mu\nu}F^{\mu\nu} + \frac{1}{2}\tilde{m}_\omega^2 \omega_\mu \omega^\mu, \tag{B.1}$$

where the tilded quantities correspond to the bare, unrenormalized parameters. Defining

$$\sigma = Z^{1/2}\sigma_r$$
$$\psi = Z_2^{1/2}\psi_r$$
$$\omega^\mu = Z_3^{1/2}\omega_r^\mu, \tag{B.2}$$

we proceed to renormalized fields. Inserting Eq. (B.2) in Eq. (B.1), the Lagrangian can be split into two terms

$$\mathcal{L} = \mathcal{L}' + \mathcal{L}_{CT}, \tag{B.3}$$

where $\mathcal{L}'$ resembles the form of the original, unrenormalized Lagrangian (B.1). To achieve this, it is convenient to invoke the following definitions

$$Z_0 = \frac{\tilde{g}_\sigma}{g_\sigma} Z_2 Z^{1/2},$$
$$Z_1 = \frac{\tilde{g}_\omega}{g_\omega} Z_2 Z_3^{1/2}. \tag{B.4}$$

Obviously, we have to assume $g_\sigma \neq 0$ and $g_\omega \neq 0$ here. Introducing the following parameters

$$\alpha_\sigma = Z - 1$$
$$\alpha_2 = m_\sigma^2 - Z\tilde{m}_\sigma^2$$
$$\alpha_\omega = Z_3 - 1$$
$$\alpha_{2\omega} = Z_3 \tilde{m}_\omega^2 - m_\omega^2$$
$$\alpha_3 = Z^{3/2} \tilde{c}_3 - c_3$$
$$\alpha_4 = Z^2 \tilde{c}_4 - c_4$$

$$\delta_2 = Z_2 - 1$$
$$\delta_m = Z_2 \tilde{m} - m$$
$$\delta_1 = 1 - Z_1$$
$$\delta_0 = Z_0 - 1, \quad \text{(B.5)}$$

we split the original Lagrangian into the two parts

$$\mathcal{L}' = \bar{\psi}_r \big( \gamma_\mu (i\partial^\mu - g_\omega \omega_r^\mu) - (m - g_\sigma \sigma_r) \big) \psi_r + \frac{1}{2} \big( \partial_\mu \sigma_r \partial^\mu \sigma_r - m_\sigma^2 \sigma_r^2 \big)$$
$$+ \frac{1}{3!} c_3 \sigma_r^3 + \frac{1}{4!} c_4 \sigma_r^4 - \frac{1}{4} (F_r^{\mu\nu})^2 + \frac{1}{2} m_\omega^2 (\omega_r^\mu)^2 , \quad \text{(B.6)}$$

and

$$\mathcal{L}_{CT} = \delta_0 g_\sigma \sigma_r (\bar{\psi}_r \psi_r) + \delta_1 g_v \omega_r^\mu (\bar{\psi}_r \gamma_\mu \psi_r) + \delta_2 (\bar{\psi}_r i\slashed{\partial} \psi_r) - \delta_m (\bar{\psi}_r \psi_r)$$
$$+ \frac{1}{2!} \alpha_2 \sigma_r^2 + \frac{1}{3!} \alpha_3 \sigma_r^3 + \frac{1}{4!} \alpha_4 \sigma_r^4 + \frac{1}{2} \alpha_\sigma \partial_\mu \sigma_r \partial^\mu \sigma_r - \frac{1}{4} \alpha_\omega (F_r^{\mu\nu})^2 + \frac{1}{2} \alpha_{2\omega} (\omega_r^\mu)^2 . \quad \text{(B.7)}$$

While Eq. (B.6) resembles the form of Eq. (B.1), all bare quantities are replaced by their renormalized, physical counter parts.

Let us focus on the counter term Lagrangian, Eq. (B.7). A coupling constant renormalization is not required, $\delta_0 \equiv \delta_1 \equiv 0$. Moreover, the momentum renormalization of the fermion field is not needed here, $\delta_2 \equiv 0$. Finally, as no mass term for a vector field can be generated dynamically in the vacuum (cf. appendix B.2), the respective mass renormalization counter term is unnecessary, $\alpha_{2\omega} \equiv 0$. Otherwise also the photon would acquire a mass in quantum electrodynamics. Using these results in Eq. (B.7), we obtain

$$\mathcal{L}_{CT} = -\delta_m \bar{\psi}_r \psi_r + \frac{1}{2!} \alpha_2 \sigma_r^2 + \frac{1}{3!} \alpha_3 \sigma_r^3 + \frac{1}{4!} \alpha_4 \sigma_r^4 + \frac{1}{2} \alpha_\sigma \partial_\mu \sigma_r \partial^\mu \sigma_r - \frac{1}{4} \alpha_\omega (F_r^{\mu\nu})^2 . \quad \text{(B.8)}$$

Eqs. (B.4) and (B.5) reduce to

$$\delta_m = \tilde{m} - m$$
$$\alpha_2 = m_\sigma^2 - (g_\sigma/\tilde{g}_\sigma)^2 \tilde{m}_\sigma^2$$
$$\alpha_3 = (g_\sigma/\tilde{g}_\sigma)^3 \tilde{c}_3 - c_3$$
$$\alpha_4 = (g_\sigma/\tilde{g}_\sigma)^4 \tilde{c}_4 - c_4$$
$$\alpha_\sigma = (g_\sigma/\tilde{g}_\sigma)^2 - 1$$

$$\alpha_\omega = \frac{m_\omega^2}{\tilde{m}_\omega^2} - 1$$
$$\frac{\tilde{g}_\omega}{\tilde{m}_\omega} = \frac{g_\omega}{m_\omega} , \quad \text{(B.9)}$$

## B.1. WALECKA MODEL WITH COUNTER TERM RENORMALIZATION

where we demanded $\tilde{m}_\omega \neq 0$ and $m_\omega \neq 0$. For the vector fields the ratio of coupling constant and mass remains unchanged, when turning from bare to physical parameters.

Conventionally, the Walecka model is dealt with a counter term Lagrangian slightly different from Eq. (B.8) (cf. [1, 35]). Instead of the counter term contribution $-\delta_m \bar{\psi}_r \psi_r$ related to the fermion mass term, one invokes the counter term contribution $\alpha_1 \sigma_r$.

Let us demonstrate that these two variants of $\mathcal{L}_{CT}$ are equivalent. To do this, it is convenient to turn to the language of Feynman diagrams.

Instead of eliminating the divergence in the self-consistent fermion mass by inserting the counter term $\delta_m$ in every fermion line, due to the scalar-fermion interaction $g_\sigma \sigma_r \bar{\psi}_r \psi_r$, one can alternatively invoke the counter term $\alpha_1$ to remove the very same divergence. Inserting the counter term $\alpha_1$ coupled via a scalar propagator with mass-squared $m_\sigma^2 - \alpha_2$ and zero momentum transfer in every fermion line (cf. Eq. (3.7)), it plays exactly the same role as $\delta_m$. We obtain the following identity

$$\delta_m = -\alpha_1 \left(-\frac{i}{m_\sigma^2}\right) i g_\sigma = -\frac{g_\sigma^2}{m_\sigma^2} \frac{\alpha_1}{g_\sigma}. \tag{B.10}$$

In result, the counter term Lagrangian can also be written as

$$\mathcal{L}_{CT} = \alpha_1 \sigma_r + \frac{1}{2!}\alpha_2 \sigma_r^2 + \frac{1}{3!}\alpha_3 \sigma_r^3 + \frac{1}{4!}\alpha_4 \sigma_r^4 + \frac{1}{2}\alpha_\sigma \partial_\mu \sigma_r \partial^\mu \sigma_r - \frac{1}{4}\alpha_\omega \left(F_r^{\mu\nu}\right)^2. \tag{B.11}$$

This is the counter term Lagrangian utilized in [36].

## B.2 Proof: "$-$" loops with $\gamma^\nu$ insertions vanish at zero external momentum transfer

In this section we show that all effective interactions due to "$-$" loops with at least one $\gamma^\nu$-insertion vanish at zero external momentum transfer. This does not mean that every single loop is zero, but rather that the linear combination of the loops corresponding to an effective interaction (with respective weights) vanishes.

Let us first show that all "$-$" loops with just one type of $\gamma^\nu$-insertions ($\nu$ fixed) and any number of 1 insertions vanish. In order to prove this, it is sufficient to show that all effective interactions descending from "$-$" loops with $\gamma^\nu$-insertions ($\nu$ fixed) only

$$\sim \int d^d p \, \text{Tr}\left\{[\gamma^\nu G_-^H(p)]^n\right\}, \quad n \in \mathbb{N} \tag{B.12}$$

vanish, as loops with $n$ $\gamma^\nu$-insertions and $l$ 1-insertions can be derived from '$n$ $\gamma^\nu$-insertions only' diagrams via differentiation for $m$

$$\sim \left(\frac{\partial}{\partial m}\right)^l \int d^d p \, \text{Tr}\left\{[\gamma^\nu G_-^H(p)]^n\right\}, \quad (n, l) \in \mathbb{N}. \tag{B.13}$$

Noting that ($\nu$ fixed, no summation)

$$[\gamma^\nu G_-^H(p+q)]^2 = \frac{1}{(p+q)^2 - m^2}\left[2(p^\nu + q^\nu)\gamma^\nu G_-^H(p+q) - g^{\nu\nu}\right], \tag{B.14}$$

and as

$$\gamma^\nu G_-^H(p+q) = \frac{\gamma^\nu(\slashed{p}+\slashed{q}+m)}{(p+q)^2 - m^2} \tag{B.15}$$

also

$$\frac{\partial}{\partial q_\nu}\left[\gamma^\nu G_-^H(p+q)\right] = -\frac{1}{(p+q)^2 - m^2}\left[2(p^\nu + q^\nu)\gamma^\nu G_-^H(p+q) - g^{\nu\nu}\right], \tag{B.16}$$

we obtain the following relation

$$\left[\gamma^\nu G_-^H(p+q)\right]^2 = -\frac{\partial}{\partial q_\nu}\left[\gamma^\nu G_-^H(p+q)\right]. \tag{B.17}$$

From

$$\mathrm{Tr}\left\{\left[\gamma^\nu G_-^H(p+q)\right]^n\right\} = -\mathrm{Tr}\left\{\left[\gamma^\nu G_-^H(p+q)\right]^{n-2}\frac{\partial}{\partial q_\nu}\left[\gamma^\nu G_-^H(p+q)\right]\right\}, \tag{B.18}$$

it is then clear that

$$\mathrm{Tr}\left\{\left[\gamma^\nu(p+q)\right]^n\right\} \sim \frac{\partial}{\partial q_\nu}\mathrm{Tr}\left\{\left[\gamma^\nu G_-^H(p+q)\right]^{n-1}\right\}. \tag{B.19}$$

Consequently, reinserting relation (B.19), we obtain

$$\mathrm{Tr}\left\{\left[\gamma^\nu G_-^H(p+q)\right]^n\right\} \sim \left(\frac{\partial}{\partial q_\nu}\right)^{n-1}\mathrm{Tr}\left\{\gamma^\nu G_-^H(p+q)\right\}, \tag{B.20}$$

which results in

$$\int d^dp\,\mathrm{Tr}\left\{\left[\gamma^\nu G_-^H(p+q)\right]^n\right\} \sim \left(\frac{\partial}{\partial q_\nu}\right)^{n-1}\int d^dp\,\mathrm{Tr}\left\{\gamma^\nu G_-^H(p+q)\right\}. \tag{B.21}$$

As

$$\int d^dp\,\mathrm{Tr}\left\{\left[\gamma^\nu G_-^H(p+q)\right]\right\} \equiv 0, \tag{B.22}$$

Eq. (B.21) in particular amounts to

$$\int d^dp\,\mathrm{Tr}\left\{\left[\gamma^\nu G_-^H(p)\right]^n\right\} = 0, \tag{B.23}$$

and the desired relation is shown.

It remains to be shown that those "$-$" loops incorporating all different kinds of insertions occurring in the Walecka model (1, $\gamma^0$, $\gamma^1$, and in the model in 3+1 dimensions also $\gamma^2$, $\gamma^3$) vanish as well. Therefore note that

$$-\frac{\partial}{\partial q_i}G_-^H(p+q) = G_-^H(p+q)\gamma^i G_-^H(p+q), \quad i=1\ldots 3. \tag{B.24}$$

This means that it should be possible to write all "−" loops with $n$ $\gamma^0$-insertions, $l$ 1-insertions and $s$ $\gamma^1$-insertions as

$$\sim \left(\frac{\partial}{\partial q_1}\right)^s \left(\frac{\partial}{\partial m}\right)^l \int \mathrm{d}^d p \, \mathrm{Tr}\left\{[\gamma^0 G_-^H(p+q)]^n\right\}\bigg|_{q=0}, \quad (n,l,s) \in \mathbb{N}. \tag{B.25}$$

In an analogous way, one can account for $\gamma^2$ as well as $\gamma^3$ insertions. As Eq. (B.25) vanishes, the proof is complete.

## B.3  Vanishing effective interactions

In this section we show that the effective interactions descending from a "−" loop with one $\gamma^\mu$ and two 1 insertions as well as that with $\gamma^\mu$, $\gamma^\nu$ and $\gamma^\sigma$ insertions vanish in $d$ dimensions, as long as the integral is handled with dimensional regularization. Since the integral is then finite, we may shift integration variables.

We demonstrate the procedure for the effective interaction arising from a "−" loop with one $\gamma^\mu$ and two 1 insertions. It is given by the sum of the following two expressions

$$\frac{iN}{(2\pi)^4} \int \mathrm{d}^d p \, \mathrm{Tr}\left\{G_-^H(p) G_-^H(p+k_1+k_2) G_-^H(p+k_1) \gamma^\mu\right\} \tag{B.26}$$

and

$$\frac{iN}{(2\pi)^4} \int \mathrm{d}^d p \, \mathrm{Tr}\left\{G_-^H(p) G_-^H(p+k_1+k_2) \gamma^\mu G_-^H(p+k_2)\right\}. \tag{B.27}$$

In order to combine Eqs. (B.26) and (B.27) it is convenient to put them on a common denominator. This can be achieved by shifting $p \to p - k_1$ in Eq. (B.26)

$$\frac{iN}{(2\pi)^4} \int \mathrm{d}^d p \, \mathrm{Tr}\left\{G_-^H(p-k_1) G_-^H(p+k_2) G_-^H(p) \gamma^\mu\right\}$$
$$= \frac{iN}{(2\pi)^4} \int \mathrm{d}^d p \, \frac{\mathrm{Tr}\left\{(\slashed{p}-\slashed{k}_1+m)(\slashed{p}+\slashed{k}_2+m)(\slashed{p}+m)\gamma^\mu\right\}}{[(p-k_1)^2-m^2][(p+k_2)^2-m^2][p^2-m^2]}, \tag{B.28}$$

and $p \to -p - k_2$ in Eq. (B.27)

$$\frac{iN}{(2\pi)^4} \int \mathrm{d}^d p \, \mathrm{Tr}\left\{G_-^H(-p-k_2) G_-^H(-p+k_1) \gamma^\mu G_-^H(-p)\right\}$$
$$= \frac{iN}{(2\pi)^4} \int \mathrm{d}^d p \, \frac{\mathrm{Tr}\left\{(-\slashed{p}-\slashed{k}_2+m)(-\slashed{p}+\slashed{k}_1+m)\gamma^\mu(-\slashed{p}+m)\right\}}{[(p-k_1)^2-m^2][(p+k_2)^2-m^2][p^2-m^2]}. \tag{B.29}$$

As the sum of the two traces in Eqs. (B.28) and (B.29) yields zero, the desired relation is shown. For the effective interaction arising from a "−" loop with $\gamma^\mu$, $\gamma^\nu$ and $\gamma^\sigma$ insertions, the procedure is the very same.

## B.4 Exact localized solutions of no-sea effective theory with four-fermion-interactions

This section is based on an approach, introduced by Lee et al. [16] to construct exact localized "classical" solutions of 1+1 dimensional field theories with four-fermion-interactions. To apply their considerations to Eq. (3.47), we have to generalize their ideas to the simultaneous appearance of scalar and vector interactions featuring two different coupling constants $G_\sigma \neq G_\omega$,

$$\mathcal{L} = \bar{\psi}\left(i\slashed{\partial} - m\right)\psi + \frac{G_\sigma^2}{2}\left(\bar{\psi}\psi\right)^2 + \frac{G_\omega^2}{2}\left(\bar{\psi}i\gamma_\nu\psi\right)\left(\bar{\psi}i\gamma^\nu\psi\right). \tag{B.30}$$

Here $\psi$ corresponds to a " + " fermion spinor with $N$ flavors. The required generalization is straight-forward. Following Lee et al., fermion bound states can be constructed. To retain analyticity, they assume that all fermions occupy one and the same energy level. Obviously, this limits their approach to fermion bound states with $n \leq N$ fermions only. It is convenient to turn to the notation of [16]. This implies the following choice of the $\gamma$-matrices

$$\gamma_0 = \sigma_3, \quad \gamma_1 = -i\sigma_1, \quad \gamma_5 = \gamma_0\gamma_1 = \sigma_2, \tag{B.31}$$

and the ansatz ($k$ denotes the flavor index)

$$\psi_k(x) = \begin{pmatrix} u_k \\ v_k \end{pmatrix} = \frac{R(x)}{\sqrt{N}}\begin{pmatrix} \cos\theta \\ \sin\theta \end{pmatrix}, \tag{B.32}$$

where $u_k$, $v_k$ have to be real valued functions ([16], below Eq. (7c)). $\theta(x)$ is determined by

$$\theta(x) = \arctan\left(\alpha\tanh\beta x\right). \tag{B.33}$$

$\alpha$ and $\beta$ are energy and mass dependent constants defined by

$$\alpha = \sqrt{\frac{m-E}{m+E}} \quad \text{and} \quad \beta = \sqrt{m^2 - E^2}. \tag{B.34}$$

By using [16], Eq. (8a), together with our Lagrangian (B.30), we infer

$$R^2(x) = \frac{2\left(E - m\cos 2\theta\right)}{G_\omega^2 - G_\sigma^2\cos^2 2\theta}. \tag{B.35}$$

Demanding the following normalization

$$n = \int_{-\infty}^{\infty} dx\, \psi^\dagger \psi = \int_{-\arctan\alpha}^{\arctan\alpha} d\theta\, \frac{-2}{G_\omega^2 - G_\sigma^2\cos^2 2\theta}, \tag{B.36}$$

we obtain

$$\tan\left(2\arctan\alpha\right) = -\frac{\sqrt{G_\omega^2 - G_\sigma^2}}{G_\omega}\tan\left(\frac{n}{2}G_\omega\sqrt{G_\omega^2 - G_\sigma^2}\right) \tag{B.37}$$

## B.4. EXACT LOCALIZED SOLUTIONS OF NO-SEA EFFECTIVE THEORY

for $G_\omega^2 > G_\sigma^2$, and

$$\tan(2\arctan\alpha) = \frac{\sqrt{G_\sigma^2 - G_\omega^2}}{G_\omega} \tanh\left(\frac{n}{2}G_\omega\sqrt{G_\sigma^2 - G_\omega^2}\right) \tag{B.38}$$

for $G_\sigma^2 > G_\omega^2$. As $\alpha \geq 0$, it is obvious that Eq. (B.37) could be fulfilled for negative $n$ only. Hence, we turn to $G_\sigma^2 > G_\omega^2$. Equation (B.38) can be rewritten as

$$2\arctan\alpha = \arctan\left[\frac{\sqrt{G_\sigma^2 - G_\omega^2}}{G_\omega}\tanh\left(\frac{n}{2}G_\omega\sqrt{G_\sigma^2 - G_\omega^2}\right)\right] \tag{B.39}$$

or

$$\frac{\alpha}{1-\alpha^2} = \frac{\sqrt{G_\sigma^2 - G_\omega^2}}{2G_\omega}\tanh\left(\frac{n}{2}G_\omega\sqrt{G_\sigma^2 - G_\omega^2}\right), \tag{B.40}$$

respectively. The "bound state mass" of $n$ fermions corresponds to the expectation value of the classical Hamiltonian. It is given by (cf. [16], Eq. (7c))

$$M_n = m\int_{-\infty}^{\infty} dx\,\bar\psi\psi = m\int_{-\arctan\alpha}^{\arctan\alpha} d\theta\,\frac{-2\cos 2\theta}{G_\omega^2 - G_\sigma^2\cos^2 2\theta}$$

$$= \frac{2m}{G_\sigma\sqrt{G_\sigma^2 - G_\omega^2}}\operatorname{arctanh}\left(\frac{G_\sigma}{\sqrt{G_\sigma^2 - G_\omega^2}}\sin(2\arctan\alpha)\right) \tag{B.41}$$

$$= \frac{2m}{G_\sigma\sqrt{G_\sigma^2 - G_\omega^2}}\operatorname{arcsinh}\left(\frac{G_\sigma}{G_\omega}\sinh\left(\frac{n}{2}G_\omega\sqrt{G_\sigma^2 - G_\omega^2}\right)\right). \tag{B.42}$$

In the last step of Eq. (B.42) we inserted Eq. (B.39). Let us investigate the stability of the $n$-fermion localized solution. To be stable, it has to be energetically favored as compared to the mass of $n$ free fermions, as well as the mass of $n$ localized fermions not interacting with each other. The first condition is true, as

$$\operatorname{Arsinh}(w\sinh z) \leq wz \quad \text{for} \quad w \geq 1, z \geq 0 \tag{B.43}$$

directly implies $M_n \leq nm$. To fulfill the latter condition, we have to ensure that $M_n \leq nM_1$. This is also true as

$$\operatorname{Arsinh}(w\sinh(nz)) \leq n\operatorname{Arsinh}(w\sinh z) \quad \text{for} \quad w \geq 1, z \geq 0. \tag{B.44}$$

In result, the $n$-fermion localized solution with mass $M_n$ is energetically favored. Finally, we determine the energy eigenvalue $E$. It is given by

$$E = m\frac{1-\alpha^2}{1+\alpha^2}. \tag{B.45}$$

Recasting Eq. (B.41) into

$$\frac{\alpha}{1+\alpha^2} = \frac{1}{2}\sin(2\arctan\alpha) = \frac{\sqrt{G_\sigma^2 - G_\omega^2}}{2G_\sigma}\tanh\left(\frac{G_\sigma\sqrt{G_\sigma^2 - G_\omega^2}}{2m}M_n\right) \quad \text{(B.46)}$$

and using Eq. (B.40), one obtains

$$E = m\frac{\cosh\left(\frac{n}{2}G_\omega\sqrt{G_\sigma^2 - G_\omega^2}\right)}{\sqrt{1 + \left(\frac{G_\sigma}{G_\omega}\right)^2 \sinh^2\left(\frac{n}{2}G_\omega\sqrt{G_\sigma^2 - G_\omega^2}\right)}}. \quad \text{(B.47)}$$

Therewith, the explicit expression for the $n$-fermion-state spinor becomes

$$\psi(x) = \frac{R(x)}{\sqrt{1 + \alpha^2\tanh^2\beta x}}\begin{pmatrix} 1 \\ \alpha\tanh\beta x \end{pmatrix}, \quad \text{(B.48)}$$

where we insert $\alpha$ and $\beta$ defined in Eq. (B.34) with $E$ from Eq. (B.47). Note again that to be a solution, $\psi(x)$ has to be real valued. Adopting the explicit values for the coupling constants, we obtain a solution of the Lagrangian (3.47), we are interested in here. For the special case $G_\sigma = G_\omega$ we refer to [16].

## B.5 Results for coefficients defined in Sec. 3.1.3

Here we collect the coefficients associated with the $n$-fermion bound state with mass $M_n$ in Sec. 3.1.3. In order to simplify the notation, we define

$$\tilde{A} = \frac{Ag_\sigma^2}{m_0^2}, \quad \tilde{B} = \frac{Bg_\omega^2}{m_0^2}, \quad c = \tilde{A} - \tilde{B}. \quad \text{(B.49)}$$

Coefficients for the scalar potential $S(x)$, Eq. (3.63),

$$s_{22} = -\frac{n^2\tilde{A}c}{4}, \quad s_{42} = \frac{n^4\tilde{A}c^2}{96}(3\tilde{A} - 5\tilde{B}),$$

$$s_{44} = \frac{n^4\tilde{A}c^2}{64}(\tilde{A} + 3\tilde{B}),$$

$$s_{52} = -\frac{n^4\tilde{A}c}{72}\left[\frac{N\tilde{A}^3}{\pi} + \frac{6m^2c}{m_0^2}\left(4A\tilde{A} - 3A\tilde{B} - B\tilde{B}\right)\right],$$

$$s_{54} = \frac{n^4\tilde{A}c}{96}\left[\frac{N\tilde{A}^2}{\pi}(3\tilde{B} - 5\tilde{A}) + \frac{12m^2c}{m_0^2}\left(5A\tilde{A} - 3A\tilde{B} - 2B\tilde{B}\right)\right],$$

$$s_{62} = -\frac{n^6\tilde{A}c^3}{23040}\left(135\tilde{A}^2 - 210\tilde{A}\tilde{B} + 91\tilde{B}^2\right),$$

$$s_{64} = -\frac{n^6\tilde{A}c^3}{3072}\left(9\tilde{A}^2 - 14\tilde{A}\tilde{B} - 27\tilde{B}^2\right),$$

$$s_{66} = -\frac{n^6\tilde{A}c^3}{1024}\left(\tilde{A}^2 + 10\tilde{A}\tilde{B} + 5\tilde{B}^2\right). \quad \text{(B.50)}$$

## B.5. RESULTS FOR COEFFICIENTS DEFINED IN SEC. 3.1.3

Coefficients for the potential $V_0(x)$, Eq. (3.64),

$$\begin{aligned}
\rho_{22} &= -\frac{n^2 \tilde{B} c}{4}, \quad \rho_{42} = -\frac{n^4 \tilde{B}^2 c^2}{48}, \\
\rho_{44} &= \frac{n^4 \tilde{B} c^2}{64}(3\tilde{A} + \tilde{B}), \\
\rho_{52} &= -\frac{n^4 \tilde{B} c}{72}\left[\frac{N\tilde{A}^3}{\pi} + \frac{6m^2 c}{m_0^2}\left(A\tilde{A} + 3\tilde{A}B - 4B\tilde{B}\right)\right], \\
\rho_{54} &= \frac{n^4 \tilde{B} c}{48}\left[-\frac{N\tilde{A}^3}{\pi} + \frac{6m^2 c}{m_0^2}\left(2A\tilde{A} + 3\tilde{A}B - 5B\tilde{B}\right)\right], \\
\rho_{62} &= -\frac{n^6 \tilde{B}^3 c^3}{1440}, \\
\rho_{64} &= -\frac{n^6 \tilde{B} c^3}{3072}\left(15\tilde{A}^2 - 42\tilde{A}\tilde{B} - 5\tilde{B}^2\right), \\
\rho_{66} &= -\frac{n^6 \tilde{B} c^3}{1024}\left(5\tilde{A}^2 + 10\tilde{A}\tilde{B} + \tilde{B}^2\right).
\end{aligned} \qquad (B.51)$$

# Bibliography

[1] J. D. Walecka, "A theory of highly condensed matter," Annals Phys. **83** (1974) 491.

[2] D. J. Gross and A. Neveu, "Dynamical symmetry breaking in asymptotically free field theories," Phys. Rev. D **10** (1974) 3235.

[3] W. Pauli, "Über den Zusammenhang des Abschlusses der Elektronengruppen im Atom mit der Komplexstruktur der Spektren," Z. Phys. **31** (1925) 765.

[4] E. Fermi, "Sulla quantizzazione del gas perfetto monoatomico," Rend. Lincei **3** (1926) 145.

[5] P. A. M. Dirac, "On the theory of quantum mechanics," Proc. Roy. Soc. Lond. A **112** (1926) 661.

[6] P. A. M. Dirac, "The quantum theory of the electron," Proc. Roy. Soc. Lond. A **117** (1928) 610.

[7] P. A. M. Dirac, "A theory of electrons and protons," Proc. Roy. Soc. Lond. A **126** (1930) 360.

[8] Y. Nambu and G. Jona-Lasinio, "Dynamical model of elementary particles based on an analogy with superconductivity. I," Phys. Rev. **122** (1961) 345.

[9] V. Schön and M. Thies, "2D model field theories at finite temperature and density," in *At the Frontier of Particle Physics: Handbook of QCD*, Boris Ioffe Festschrift, ed. M. Shifman, vol. 3, ch. 33, p. 1945, World Scientific, Singapore (2001) [arXiv:hep-th/0008175].

[10] J. Feinberg, "All about the static fermion bags in the Gross-Neveu model," Annals Phys. **309** (2004) 166 [arXiv:hep-th/0305240].

[11] M. Thies, "From relativistic quantum fields to condensed matter and back again: Updating the Gross-Neveu phase diagram," J. Phys. A **39** (2006) 12707 [arXiv:hep-th/0601049].

[12] R. F. Dashen, B. Hasslacher and A. Neveu, "Semiclassical bound states in an asymptotically free theory," Phys. Rev. D **12** (1975) 2443.

[13] S. S. Shei, "Semiclassical bound states in a model with chiral symmetry," Phys. Rev. D **14** (1976) 535.

[14] R. Pausch, M. Thies and V. L. Dolman, "Solving the Gross-Neveu model with relativistic many body methods," Z. Phys. A **338** (1991) 441.

[15] F. Karbstein and M. Thies, "Divergence of the axial current and fermion density in Gross-Neveu models," Phys. Rev. D **76** (2007) 085009 [arXiv:0706.0424 [hep-th]].

[16] S. Y. Lee, T. K. Kuo and A. Gavrielides, "The exact localized solutions of two-dimensional field theories of massice fermions with fermi interactions," Phys. Rev. D **12** (1975) 2249.

[17] S. Y. Lee and A. Gavrielides, "Quantization of the 'bound' state solutions in two-dimensional field theories of massive fermions," Phys. Rev. D **12** (1975) 3880.

[18] E. J. Levinson and D. H. Boal, "Selfenergy corrections to fermions in the presence of a thermal background," Phys. Rev. D **31** (1985) 3280.

[19] K. Ahmed and S. S. Masood, "Finite temperature and density renormalization effects in QED," Phys. Rev. D **35** (1987) 4020.

[20] M. Thies and K. Urlichs, "From non-degenerate conducting polymers to dense matter in the massive Gross-Neveu model," Phys. Rev. D **72** (2005) 105008, Appendix B. [arXiv:hep-th/0505024].

[21] J. Feinberg and A. Zee, "Fermion bags in the massive Gross-Neveu model," Phys. Lett. B **411** (1997) 134 [arXiv:hep-th/9610009].

[22] M. Thies and K. Urlichs, "Baryons in massive Gross-Neveu models," Phys. Rev. D **71** (2005) 105008 [arXiv:hep-th/0502210].

[23] J. Feinberg and S. Hillel, "Stable fermion bag solitons in the massive Gross-Neveu model: Inverse scattering analysis," Phys. Rev. D **72** (2005) 105009 [arXiv:hep-th/0509019].

[24] M. Gell-Mann, R. J. Oakes and B. Renner, "Behavior of current divergences under $SU(3) \times SU(3)$," Phys. Rev. **175** (1968) 2195.

[25] S. Weinberg, "Pion scattering lengths," Phys. Rev. Lett. **17** (1966) 616.

[26] S. R. Coleman, "There are no Goldstone bosons in two-dimensions," Commun. Math. Phys. **31** (1973) 259.

[27] E. Farhi, N. Graham, R. L. Jaffe and H. Weigel, "Heavy fermion stabilization of solitons in 1+1 dimensions," Nucl. Phys. B **585** (2000) 443 [arXiv:hep-th/0003144].

[28] J. Goldstone and F. Wilczek, "Fractional quantum numbers on solitons," Phys. Rev. Lett. **47** (1981) 986.

[29] R. Jackiw and G. W. Semenoff, "Continuum Quantum Field Theory for a linearly conjugated diatomic polymer With Fermion Fractionization," Phys. Rev. Lett. **50** (1983) 439.

[30] M. Thies and K. Ohta, "Continuum light cone quantization of Gross-Neveu models," Phys. Rev. D **48** (1993) 5883.

[31] C. Boehmer, F. Karbstein and M. Thies, "Baryons in the large N limit of the massive two-dimensional Nambu-Jona-Lasinio model," Phys. Rev. D **77** (2008) 125031 [arXiv:0803.1369 [hep-th]].

[32] C. Boehmer, M. Thies and K. Urlichs, "Tricritical behavior of the massive chiral Gross-Neveu model," Phys. Rev. D **75** (2007) 105017 [arXiv:hep-th/0702201].

[33] C. Boehmer, U. Fritsch, S. Kraus and M. Thies, "Phase structure of the massive chiral Gross-Neveu model from Hartree-Fock," Phys. Rev. D **78** (2008) 065043 [arXiv:0807.2571 [hep-th]].

[34] K. Urlichs, "Baryons and baryonic matter in four-fermion interaction models," Ph.D. thesis, Erlangen, 2007

[35] B. D. Serot and J. D. Walecka, "Recent progress in quantum hadrodynamics," Int. J. Mod. Phys. E **6** (1997) 515 [arXiv:nucl-th/9701058].

[36] B. D. Serot and J. D. Walecka, "The relativistic nuclear many body problem," In: Advances in nuclear physics, J. W. Negele and E. Vogt (eds.), Vol. 16. New York: Plenum Press (1986).

[37] T. C. Ferree, C. E. Price and J. R. Shepard, "Exact one-loop vacuum polarization effects in (1+1)-dimensional quantum hydrodynamics," Phys. Rev. C **47** (1993) 573.

[38] M. E. Peskin and D. V. Schroeder, "An Introduction To Quantum Field Theory," Reading, USA: Addison-Wesley (1995).

[39] F. Karbstein and M. Thies, "Integrating out the Dirac sea: Effective field theory approach to exactly solvable four-fermion models," Phys. Rev. D **77** (2008) 025008 [arXiv:0708.3176 [hep-th]].

Die VDM Verlagsservicegesellschaft sucht für wissenschaftliche Verlage abgeschlossene und herausragende

# Dissertationen, Habilitationen, Diplomarbeiten, Master Theses, Magisterarbeiten usw.

## für die kostenlose Publikation als Fachbuch.

Sie verfügen über eine Arbeit, die hohen inhaltlichen und formalen Ansprüchen genügt, und haben Interesse an einer honorarvergüteten Publikation?

Dann senden Sie bitte erste Informationen über sich und Ihre Arbeit per Email an *info@vdm-vsg.de*.

## Sie erhalten kurzfristig unser Feedback!

VDM Verlagsservicegesellschaft mbH
Dudweiler Landstr. 99
D - 66123 Saarbrücken

Telefon +49 681 3720 174
Fax +49 681 3720 1749

**www.vdm-vsg.de**

Die VDM Verlagsservicegesellschaft mbH vertritt

Printed by Books on Demand GmbH, Norderstedt / Germany